Residual Stresses in Friction Stir Welding

Residual Stresses in Friction Stir Welding

Nilesh Kumar
Rajiv S. Mishra
John A. Baumann

AMSTERDAM • BOSTON • HEIDELBERG • LONDON
NEW YORK • OXFORD • PARIS • SAN DIEGO
SAN FRANCISCO • SINGAPORE • SYDNEY • TOKYO
Butterworth-Heinemann is an imprint of Elsevier

ELSEVIER

Acquiring Editor: Steve Merken
Editorial Project Manager: Jeff Freeland
Project Manager: Mohana Natarajan

Butterworth-Heinemann is an imprint of Elsevier
225 Wyman Street, Waltham, MA 02451, USA
The Boulevard, Langford Lane, Kidlington, Oxford, OX5 1GB, UK

First published 2014

British Library Cataloguing-in-Publication Data
A catalogue record for this book is available from the British Library

Library of Congress Cataloging-in-Publication Data
A catalog record for this book is available from the Library of Congress

ISBN: 978-0-12-800150-9

For information on all Butterworth-Heinemann publications
visit our website at **store.elsevier.com**

This book has been manufactured using Print On Demand technology. Each copy is produced to order and is limited to black ink. The online version of this book will show color figures where appropriate.

Working together
to grow libraries in
developing countries

www.elsevier.com • www.bookaid.org

CONTENTS

Preface to the Friction Stir Welding and Processing Book Series

Twenty-two years since its invention, friction stir welding has been established as a significant solid-state welding technique. Its application spans across all sectors; aerospace, automotive, railways, shipbuilding, to name a few. The friction stir process leads to many unique microstructural features and that has been the basis for development of a broader metallurgical tool of friction stir processing for microstructural modification. In the last 15 years, the burgeoning number of technical papers shows maturing of the field. Additionally, a few edited books, review papers, and conference proceedings provide a good survey of the progress made. With the volume of available information increasing, a need was felt to have a short format book series to bring out critical reviews dedicated to the overall field of friction stir welding and processing. As the founding series editor, it is my intention to bring a number of topical volumes each year to further the science and technology of friction stir welding and processing.

Stephen R. Merken, Acquisitions Editor at Elsevier, has been instrumental in launch of this series and discussion with him has shaped its format. We look forward to the growth of this series and hope that it will serve as a seminal source of information for researchers engaged in the field of friction stir welding and processing. It will be equally valuable for engineers and researchers engaged in advanced and innovative manufacturing techniques.

Rajiv S. Mishra
University of North Texas
October 18, 2013

CHAPTER 1

Introduction

Today welding is being widely used as a joining process in myriad of structural applications such as buildings, offshore petroleum rigs, pressure vessels, and ships. In fact, before World War II, all the ships contained riveted structures, whereas today they have been replaced mostly with welding. The reason for such a widespread use is several advantages that welding offers over competing joining techniques, especially riveting. The welded structures are generally lighter and have higher joint efficiency. It also simplifies the joint design which results in smaller fabrication time. Despite possessing several unique joining characteristics, it has several drawbacks. In fusion welding techniques, defects such as liquation cracking, solidification induced porosities, and inclusions from slag are quite common. Residual stresses (RS) are common to both fusion and solid-state welding techniques. RS are the result of sharp inhomogeneous thermal gradient which results in complex thermal stresses in the weld. It eventually leads to the introduction of RS and resulting distortion of the weldments.

The scope of this book is to get an insight into RS induced in components joined by friction stir welding (FSW) technique by summarizing published literature. It will help practitioners of this technology to develop understanding about the factors controlling RS generation and their mitigation.

1.1 RESIDUAL STRESSES

Residual stresses are defined as self-equilibrating stresses which exist in an elastic body even in absence of external loads (thermal and/or mechanical). In Figure 1.1, the origin of RS has been depicted, schematically, as a consequence of interactions among thermal history (time and temperature), deformation (stress and strain), and material microstructure [1]. Clearly, these stresses are results of thermal and/or mechanical treatment of components during their manufacturing

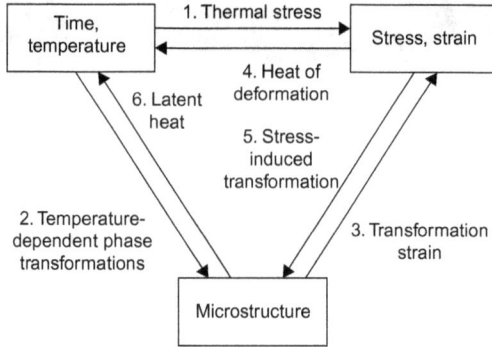

Figure 1.1 Schematic representation of residual stress as a consequence of interactions among factors as shown in boxes [1].

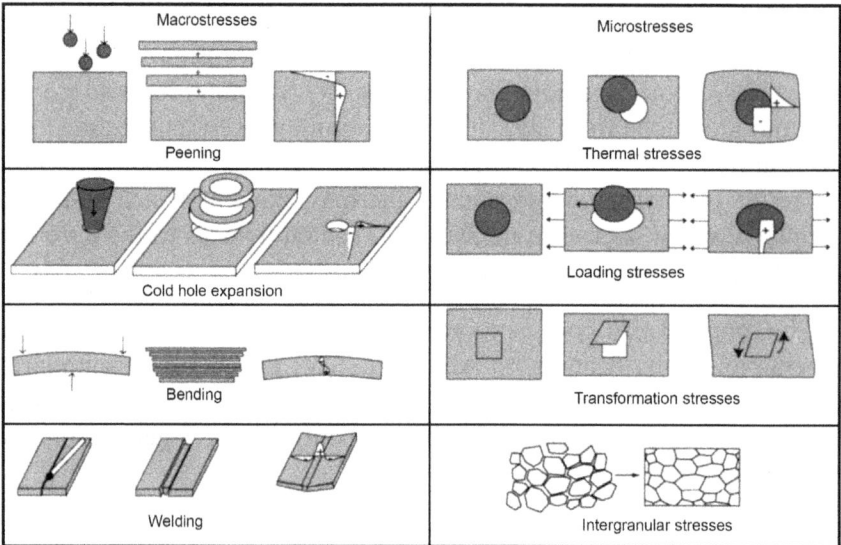

Figure 1.2 Schematic representation of RS as a misfit between different regions under different thermo-mechanical processing conditions [2].

(through casting, metalworking, welding, physical vapor deposition, etc.). Figure 1.2 illustrates some of the manufacturing scenarios which can introduce RS in various components [2]. The complete removal of RS may not be possible. However, by following proper manufacturing techniques, they can be mitigated to a large extent. Hence, the knowledge of factors influencing the state of RS is very important for managing RS.

RS occur at different length scales in materials. Depending upon length scale at which they are present, they can be classified into type I, II, or III. Type I is referred to as macrostress and varies over the scale of the structure. Type II varies and self-equilibrates over the scale of several grains. Type III varies over less than the scale of a grain. The stresses existing in welded component due to nonuniform heating and cooling and sharp thermal gradient can be categorized into type I. Stresses existing in composite materials due to different elastic and thermal properties can come under type II. The stresses associated with dislocations and vacancies come under type III. All these three types of RS have been summarized in Figure 1.3 [2]. In a given component depending on its processing history, all three types of stresses may coexist. Since the present book deals with RS in weldments, it will focus only on type I RS.

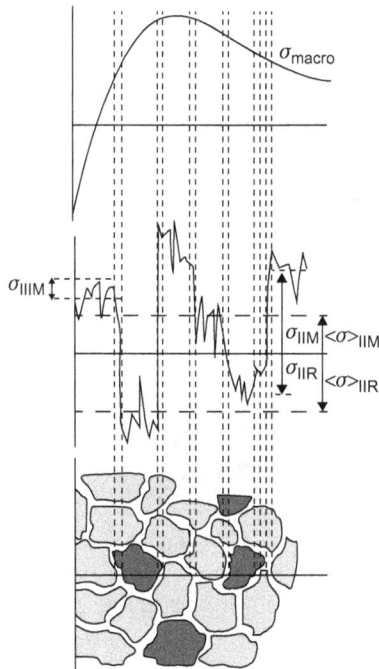

M and R denote matrix and reinforcement respectively

Figure 1.3 Schematic representation of type I, II, and III RS [2].

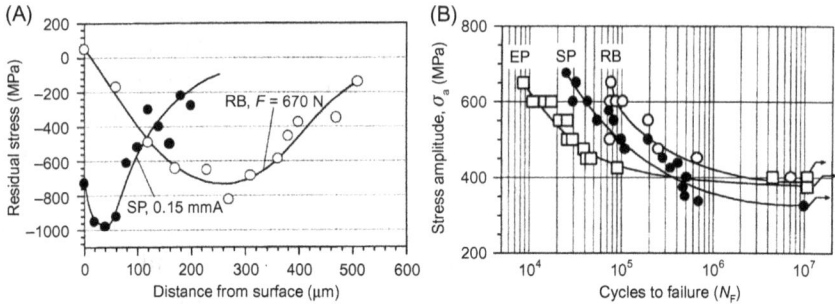

Figure 1.4 Effect of surface treatment techniques: (A) shot peening (SP) and roller-burnishing (RB) on the state of RS and (B) on fatigue life (EP: electrolytic polishing) [4].

1.2 IMPLICATION OF RS

Depending upon the nature of RS, the presence of RS can have detrimental or beneficial effects on the service life or performance of a component. For example, surface treatment techniques such as shot peening which introduces compressive RS are known to improve fatigue life of the parts. Cold hole expansion (Figure 1.2) is another example which introduces compressive stresses around the hole thereby improving fatigue life [3]. Figure 1.4 has been included to illustrate the positive effect of such surface treatments (shot peening, roller-burnishing, etc.) on mechanical properties, such as fatigue [4]. In Figure 1.4A, it can be noted that these techniques introduce compressive stresses in the surface of the components. Compressive RS help improve the fatigue life by retarding the rate of surface/subsurface crack growth. The toughening of glasses to improve fracture strength by way of introducing compressive stresses on the surface is another example of beneficial effect of RS.

Due to detrimental effect of RS (compressive or tensile), structural components may get severely distorted which can result in the rejection of the parts or the presence of RS can bring down the service life of a component quite significantly.

Hence, the knowledge of RS as a result of primary (e.g., casting) or secondary (e.g., machining) manufacturing processes is essential. As mentioned earlier, the scope of the book is limited to RS in FSW process, so in the following sections, various factors governing RS, the role of RS on mechanical properties, and how they can be reduced or eliminated is discussed.

CHAPTER 2

A Brief Introduction to FSW

FSW was developed at The Welding Institute (TWI) in 1991 [5]. The working principle of FSW/P is remarkably simple [6]. In this process, in its simplest form, a nonconsumable rotating tool plunges inside a material at the joint line of two butting plates to be welded and traverses along the weld line, which results in weld formation and modification of microstructure. During welding, the temperature in the material never exceeds the solidus temperature of the material and the whole process is performed in the solid state. The schematic of the process has been shown in Figure 2.1 and relevant regions of FSW process have been labeled. Table 2.1 lists important beneficial effects of FSW technique [7].

Due to similarity of FSW to other thermo-mechanical processes to some extent, this technique of material joining/processing is expected to lead to introduction of RS in the fabricated components. Since entire welding takes place below melting point of the alloy(s) being joined, at early stages of FSW development, the RS were thought to

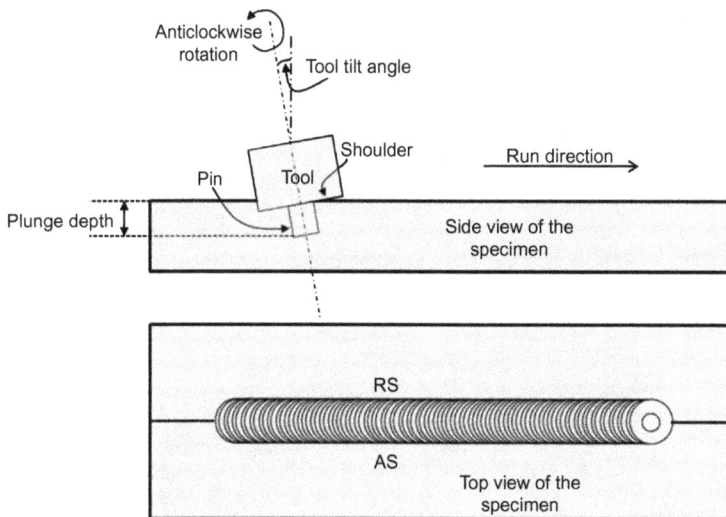

Figure 2.1 A schematic of friction stir welding process.

Table 2.1 Major Advantages of FSW Technique

Metallurgical Benefits	Environmental Benefits	Energy Benefits
Solid-state process Good dimensional stability and repeatability No loss of alloying elements Excellent metallurgical properties in the joint area Fine microstructure Absence of cracking (hot cracking) Replace multiple parts joined by fasteners	No shielding gas required No surface cleaning required Eliminate grinding wastes Eliminate solvents required for degreasing Consumable materials (such as rugs, wire, or any other gases) savings	Improved materials use (e.g., joining different thickness) allows reduction in weight) Only 2.5% of the energy needed for laser weld Decreased fuel consumption in light aircraft, automotive, and ship applications

Source: Reproduced from Ref. [7].

be negligible compared to fusion welding technique. However, the research carried out in last one decade or so has revealed that the RS in friction stir weldments can be significant. Hence, subsequent sections deal with different aspects of RS in FSW process.

CHAPTER *3*

RS in FSW Process

3.1 MATERIALS STUDIED

The materials investigated for the purpose of studying RS introduced by FSW are summarized in Figure 3.1. As can be noted, the materials explored are both ferrous and nonferrous. However, the majority of the materials investigated are aluminum-based alloys. In aluminum alloys, the precipitation strengthened alloys (2XXX, 6XXX, and 7XXX) have been studied extensively. The reason for obvious focus on precipitation strengthened aluminum alloys is related to their extensive use in automobile and aerospace industries. In nonferrous series, very little work has been done toward understanding RS in Mg and Ti based. RS in ferrous alloys also have not received much attention. Part of the reason for it is difficulty associated with welding high-temperature Ti- and Fe-based alloys. FSW community is still trying to develop wear-resistant tools capable of welding high-temperature alloys.

3.2 RS IN ALUMINUM ALLOYS

Table 3.1 summarizes maximum stresses and stresses at weld centerline for various Al alloys. The corresponding plate dimensions and processing parameters are also provided. Survey of Table 3.1 suggests that the nature of these stresses in the welded region is mostly tensile. The magnitude of RS can be as high as yield strength (YS) of the material. Moreover, it is highly dependent on the YS of the parent material. In general, high-strength materials possess higher level of RS. The 2XXX, 6XXX, and 7XXX alloys are high-strength Al alloys and on an average they demonstrate same level of RS but higher than the 5XXX series Al alloy. Generally, the 2XXX and 7XXX alloys have higher YS than 6XXX alloys. Hence, the level of RS in the 2XXX and 7XXX alloys should be higher than that of the 6XXX alloys. In fact, this is the case as per Table 2.1. However, the differences are not much. Another point which is worth discussing here is the relative magnitude of RS compared to YS of the material. Comparison is generally made with the YS of base material (BM). For example,

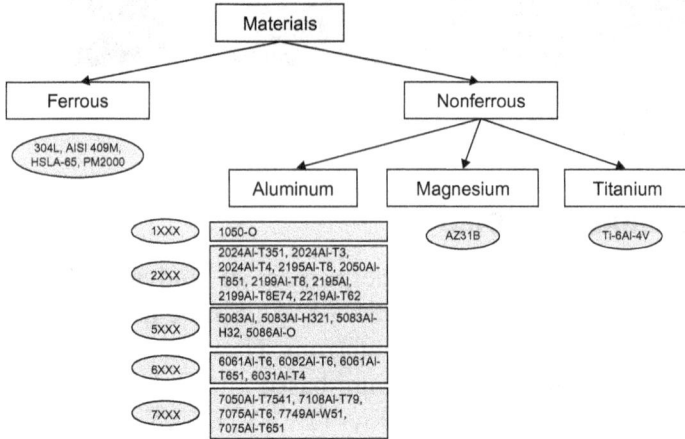

Figure 3.1 List of the materials investigated for residual stress study and its impact on various aspect of the alloy.

Lemmen et al. [21] evaluated the RS of friction stir (FS) welded 7075Al-T6 alloy. The maximum tensile RS in the workpiece was found to be 225 MPa. The YS of the BM was ~525 MPa. If comparison is made with the YS of BM, the RS is ~43% of the YS of BM. However, one should bear in mind that FSW more often than not alters the YS of heat affected zone (HAZ), thermo-mechanically affected zone (TMAZ), and nugget (dynamically recrystallized zone). For 7075Al-T6 processed by Lemmen et al. [21], the minimum YS was found to be ~325 MPa in HAZ. Comparison of RS with the YS in HAZ suggests that the magnitude is ~62 %. But such comparisons are not easy to make because of the lack of YS data of the welded component in the published literature on RS in FSW.

3.2.1 1XXX Series Aluminum Alloy
The 1XXX series aluminum alloys are not structural materials (mostly used in chemical industry). Hence, there is hardly any RS study exclusively on the 1XXX series aluminum alloys. However, Fratini and Zuccarello [32] studied RS in 1050Al alloy in O-tempered condition along with other aluminum alloys for comparison purposes. Residual stress very close to the YS of the 1050Al−O alloy (~103 MPa) was reported along the depth. It should be noted that this is maximum among tensile and compressive stresses measured in the welds made using three different levels of heat input. No particular correlation between processing parameters used and RS measured for this alloy was obtained in that study.

Table 3.1 Summary of Residual Stress Magnitude, Nature, and Characterization Technique Used Along FSW Process Parameters

Alloy	YS (MPa)	Tensile RS (Max) (MPa)	Tensile RS (Weld Centerline) (MPa)	Tool Rotation Rate (rpm)	Tool Traverse Speed (mm/min)	Plate Dimension (After Joining), mm			Measurement Technique	Reference
						L	W	t		
2219Al-T62		170	110	300	60	–	–	12	Hole drilling strain gage method	[8]
		130	90	400	60	–	–	12		
		120	70	500	60	–	–	12		
		150	110	400	100	–	–	12		
		135	90	400	80	–	–	12		
2024Al-T4	325	65	25	750	150	250	100	5	Neutron diffraction	[9,10]
		90	– 25 (compressive)	750	475	250	100	5		
2024Al-T3		140	95	1500	100	1000	240	1.6	Slot sectioning stress relaxation method	[11]
2199Al-T8E74	400	150	140	800	200	250	150	5	Synchrotron diffraction	[12]
		170	160	800	300	250	150	5		
		210	170	800	400	250	150	5		
2195Al-T8	580	120	90			1000		12.7	Neutron diffraction	[13]
2024Al-T3	345	185	150	800	200	500	700	5	Neutron and Synchrotron diffraction	[14]
2199Al-T8	400	220	170	800	400	500	700	5	Neutron and Synchrotron diffraction	[14]
2098Al		275	163	1200	500	880	600	3	Synchrotron diffraction	[15]

(Continued)

Table 3.1 (Continued)

Alloy	YS (MPa)	Tensile RS (Max) (MPa)	Tensile RS (Weld Centerline) (MPa)	Tool Traverse Speed (mm/min)	Tool Rotation Rate (rpm)	Plate Dimension (After Joining), mm			Measurement Technique	Reference
						L	W	t		
2024Al-T351	470	56	40	71.5	750	100	140	3	Cut-compliance technique	[16]
2195Al	215 (weld plate)	150	100	150	300	910	300	12.5	XRD	[17]
2050Al-T851	500 (BM); 215	130	90	200	280			15	Cut-compliance technique	[18]
2024Al-T4		165	110						Hole drilling strain gage method	[19]
2024Al-T3	322	125	100	198	576	375	250	3.2	Synchrotron diffraction	[20]
2024Al-T351	318	135	75	96	348	375	250	6.3		
2024Al-T3	322	185	100	198	348	1135	250	3.2		
2024Al-T3	330 (BM); 320 (Nugget)	287	200	350	550			2.5	XRD	[21]
5083Al-H19	392	60	15	200		150	200	3	Synchrotron diffraction	[22]
5083Al-H321	263	30	10	80	500	500	1000	8	Synchrotron diffraction	[23]
5083Al-H321		100	50	185	348			6	Synchrotron diffraction	[24]
5083Al-H321	228	155	120	160	600	1000	200	3.5	Slot sectioning stress relaxation method	[11]
6061Al-T6	276	140	50	279		300	300	6	Neutron diffraction	[25]
	276	200	105	787		300	300	6		

Material									Technique	Ref.
6082Al-T6		150	50	300	560	150	120	3	Synchrotron diffraction	[26]
6061Al-T6		120	60	279	1250	306	306	6.5	Neutron diffraction	[27]
6013Al-T4	203	200	80	1000	1500			1.8	XRD	[21]
7075Al-T6	510	92	40	100	715				Cut-compliance technique	[28]
7449Al		200	200	250	225	1000	600	12	Synchrotron diffraction	[15]
7075Al-T6	525 (BM); 390 (Nugget)	225	200	300	280			2	XRD	[21]
7108Al-T79	400	200	120	600					Ultrasonic wave technique	[29]
7050Al-T7541		195	100	408	540	914	204	6.4	Cut-compliance technique	[30]
7749Al	583	230	210	250	225	1000	150	12.2	Synchrotron diffraction	[31]

3.2.2 2XXX Series Aluminum Alloy

The 2XXX series aluminum alloys are used very extensively in automobile and aerospace industries due to their high strength to weight ratio. Owing to their popularity in these industries, researchers in FSW community have studied RS in the 2XXX series weldments since the FSW technology was in its early stages of development [33]. All the 2XXX series aluminum alloys studied for this purpose so far are listed in Figure 3.1. In the 2XXX series aluminum alloys, 2024Al have received widespread attention from the researchers. Although 2050Al, 2195Al, and 2199Al contain Cu as the major alloying element, they have been termed as Al−Li alloys. Due to their promise for significant weight reduction in aerospace structures, these alloys have also been explored to a great extent.

Maximum tensile RS are mostly longitudinal RS which ranged from 12% [16] to 87% [21] of their YS at room temperature. Interestingly, the majority of the maximum compressive stresses in the weldments are also longitudinal RS. It is quite conceivable as the net RS should be zero in any component. Hence high tensile stresses are balanced by high compressive RS of same directionality. In the nugget, the longitudinal RS also constitute 8% [9] to 61% [21] of the room temperature YS and they are mostly tensile in nature. The compressive RS are mostly transverse stresses in the nugget. On an average, transverse RS are smaller than longitudinal RS in terms of magnitude. It should be noted here that relative comparison of RS is being made with YS of BM at room temperature.

The majority of the work on the RS in the weldments shows M-shaped distribution of longitudinal stress in the region across the weld centerline. The maximum in RS occurred just outside the shoulder of the tool. However, some researchers have reported even inverted V-shaped distribution of longitudinal RS [9]. The distribution shown by transverse residual stress is almost flat across the weld centerline with slight bulge either toward tensile or compressive side mostly at weld centerline.

3.2.3 5XXX Series Aluminum Alloys

The trends shown in 5XXX series aluminum alloys by longitudinal RS (M-shaped curves) are same as the ones observed for 2024Al alloys. However, the trend shown by transverse component of RS for this series of alloys, in the majority of the cases, are different from those observed for 2024Al alloys. For the 5XXX series aluminum alloys, the trends for longitudinal and transverse RS are similar, that is, transverse RS also exhibit M-shaped distribution of the stresses in the weldments [11,22−24,34−36]. The maximum tensile longitudinal RS range from 11% [23] to 68% [11] of their room temperature YS. The tensile stresses at weld centerline ranged from 4% [23] to 53% [35] of YS of BM at room temperature. Here also in most of the cases, maximum compressive stresses are longitudinal RS except the work reported in Refs. [22] and [35]. In the nugget, again, the majority of the published longitudinal and transverse RS are tensile in nature for this series of alloy.

3.2.4 6XXX Series Aluminum Alloys

Due to very good corrosion resistance, weldability, and formability, these alloys are extensively used in automobile and construction industries. The majority of the residual stress studies have focused on 6061Al in T6 temper condition with the exception of 6013Al-T4 and 6082Al-T6 alloys. As reported for the 2XXX and 5XXX series aluminum alloys, maximum tensile stresses were longitudinal RS which ranged from 43% [27] to 99% [21] of the YS of the BM at room temperature. The highest compressive stresses were also found to be longitudinal RS. In the nugget, the maximum tensile stresses were longitudinal RS (up to 53% [25] of the YS of BM at room temperature). The trend of the distribution of the individual components of RS for the 6XXX series aluminum alloys was similar to those observed for the 2XXX and 5XXX series aluminum alloys.

3.2.5 7XXX Series Aluminum Alloys

As can be seen in Figure 3.1, wide varieties of the 7XXX series aluminum alloys have been subjected to residual stress studies. The 7XXX alloys are also extensively used in aerospace, military, marine, automobile, and construction industries. Longitudinal RS are the highest tensile and compressive stresses reported in the weldments for these alloys as well. Longitudinal tensile stresses ranged from 18% [28] to 50% [21] of the YS of the BM at room temperature. Again, in the nugget, highest tensile stresses are longitudinal RS, and values as high as 38% [21] of the YS of the BM at room temperature have been reported in the literature. The trend of distribution of various components of RS is similar to those observed for aluminum alloys discussed before.

3.2.6 RS in Dissimilar Al Alloys Welds

Even in dissimilar FS welds of aluminum alloys, trends and values of RS are quite similar to those reported for various aluminum alloys [26,37–39]. However, Steuwer et al. [26] reported a discontinuity (very sharp gradient) in the magnitude of longitudinal stresses at the weld centerline. For example, the FS weld at 840 rpm and 100 mm/min showed a longitudinal residual stress of ~50 and ~100 MPa on 6082Al and 5083Al sides, respectively. It may be a result of sharp gradient in the YS across the weld centerline. Such gradient may exist if mixing between dissimilar materials is insufficient. It should be noted that the RS characterization was carried out using neutron and synchrotron diffraction as well. However, neutron diffraction did not show any such type of distribution of RS in the weld region. It may be due to low spatial resolution of neutron diffraction compared to synchrotron diffraction technique.

3.3 RS IN OTHER MATERIALS

Other nonferrous and ferrous materials have not been as extensively researched as aluminum-based alloys. However, limited research on some nonferrous materials such as AZ31B alloys [40,41] (magnesium-based alloy) and Ti−6Al−4V alloys [42,43] (titanium-based alloy) shows results quite similar to the ones reported for aluminum alloys.

The outcome of the research on residual stress of ferrous materials (mainly steels) is in line with the trend reported for aluminum alloys [44−47]. But values as high as 450 MPa (longitudinal RS) have been reported, about two times higher than the maximum tensile RS value reported for aluminum alloys [46].

3.4 SUMMARY OF RS IN FERROUS AND NONFERROUS MATERIALS

Based on the observations made on the state of RS in the weldments of different ferrous and nonferrous metals and alloys, the trend shown by RS can be schematically represented as shown in Figure 3.2. It shows two different distributions of RS across the weld in similar (Figure 3.2A) and dissimilar material welding (Figure 3.2B) conditions. Following conclusions can be made on RS:

a. In general, longitudinal stresses are tensile in nature in the welded region. In the area away from the welded region (nugget), these stresses become compressive to balance the tensile stresses.

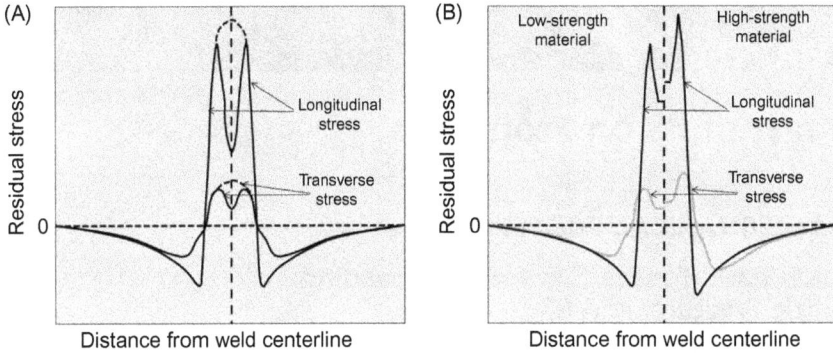

Figure 3.2 Schematic representation of the distribution of RS in (A) similar and (B) dissimilar metal conditions across the weld centerline. In (B) the discontinuity along the stress axis means a sharp stress gradient but not a real discontinuity.

b. The maximum in longitudinal tensile stresses occur either in TMAZ or HAZ region just outside the nugget (dynamically recrystallized region).

c. In general, transverse RS are also tensile in nature. However, the magnitudes of these stresses are significantly lower than that of longitudinal RS.

d. In fusion-welded plates and sheets, similar distribution and trends are shown by RS. As per Masubuchi and Martin [48], the longitudinal RS can be approximated by following mathematical expression:

$$\sigma_x(y) = \sigma_m \left\{ 1 - \left(\frac{y}{b}\right)^2 \right\} \exp\left\{ -\frac{1}{2} \left(\frac{y}{b}\right)^2 \right\} \tag{3.1}$$

where σ_m, y, and b are peak tensile longitudinal RS, distance along y coordinate, and width of the tensile residual stress zone, respectively. Given the similarity of RS curves between fusion and FS welding, Eq. (3.1) is expected to hold good for FS-welded structure also.

e. For dissimilar metal welding, these RS show discontinuity at the weld centerline. It is due to different level of the YS of the materials being joined and may be due to insufficient mixing in the weld nugget.

In some cases, maximum in longitudinal and tensile stresses have been reported at the weld centerline (as shown schematically by dotted lines in Figure 3.2A). It can be the case where compressive plastic strains are not lowered by tensile plastic strains during cooling and clamp removal of the weld.

CHAPTER 4

Effect of RS on Properties

4.1 MECHANICAL PROPERTIES

4.1.1 Static Tensile/Compressive Loading, Brittle Fracture, and RS

It is important to gain insight of what happens when a weld panel possessing residual stress is subjected to various types of loadings. Figure 3.2 schematically shows a butt-welded plate under uniform tensile loading. Curve 0 shows distribution of RS in as-welded condition [49]. Curves 1, 2, and 3 are distribution of RS when uniform tensile loading is σ_1, σ_2, and σ_3 ($\sigma_1 < \sigma_2 < \sigma_3$). Curves 1', 2', and 3' are the distribution of RS after removal of tensile loading. Following points can be inferred from this illustration:

a. Tensile region of RS are least affected with such loading. The RS present in regions away from the welded region are significantly affected.

b. As loading is increased, stress distribution in the welded panel evens out. If loading is such that the entire plate yields, in that case stress distribution becomes uniform throughout the plate and beyond this point the RS do not play any role on the stress distribution in the plate.

c. It also emphasizes to the fact that on subjecting such welded structures to repeated cyclic loading, the RS should decrease with passage of time. Section 4.1.2 provides further commentary on the influence of RS on fatigue fracture behavior.

d. RS influence only those phenomena which take place under low loading conditions as is the case with brittle fracture, stress-corrosion cracking (SCC), and even very high cycle fatigue (Figure 4.1).

With regard to brittle fracture mentioned in point (d) above, it is worthwhile to mention the results of experiments carried out by Kihara and Masubuchi [50] to understand the role of sharp notch in welded structure in the presence of RS. Figure 4.2 shows the effect of the presence of RS on fracture behavior. The fracture behavior of the

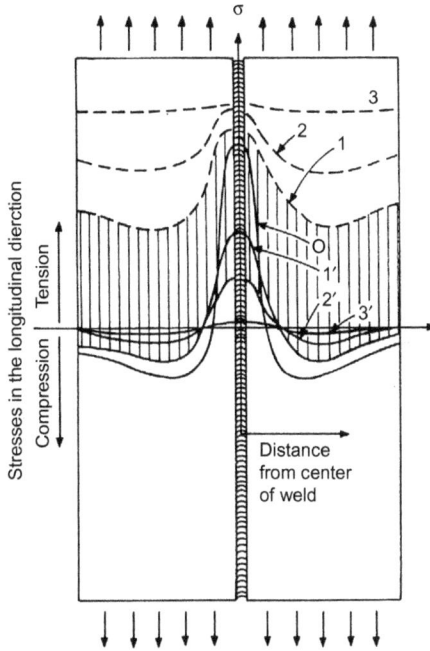

Figure 4.1 *Effect of tensile loading on residual stress distribution in a weldment joined in butt configuration (refer text for the explanation of the labels) [49].*

plate containing sharp notch is given either by curve PQR or PQST. Curve PQR describes a situation where there are no notch and RS present in the plate. Curve PQST corresponds to a situation where a notch containing specimen is tested at different temperatures in absence of residual stress. Temperatures T_f and T_a are defined as fracture transition temperature and crack transition temperature, respectively. Depending on whether temperature is below or above T_f, the fracture behavior can change from cleavage to shear. In the presence of RS, the fracture behavior is governed by the extent of loading and testing temperature. Like before, above T_f, fracture is governed by ultimate tensile strength of the material. The crack may be initiated if test temperature is between T_a and T_f; however, it eventually gets arrested. In case testing temperature is below T_a, depending upon applied load whether below or above dashed line VW, partial or complete fracture of the plate would occur. The plot shown in Figure 4.2 was drawn by Kihara and Masubuchi [50] based on their experiments on welded and notched steel plates. Figure 4.2 is a generalized representation of their experimental findings.

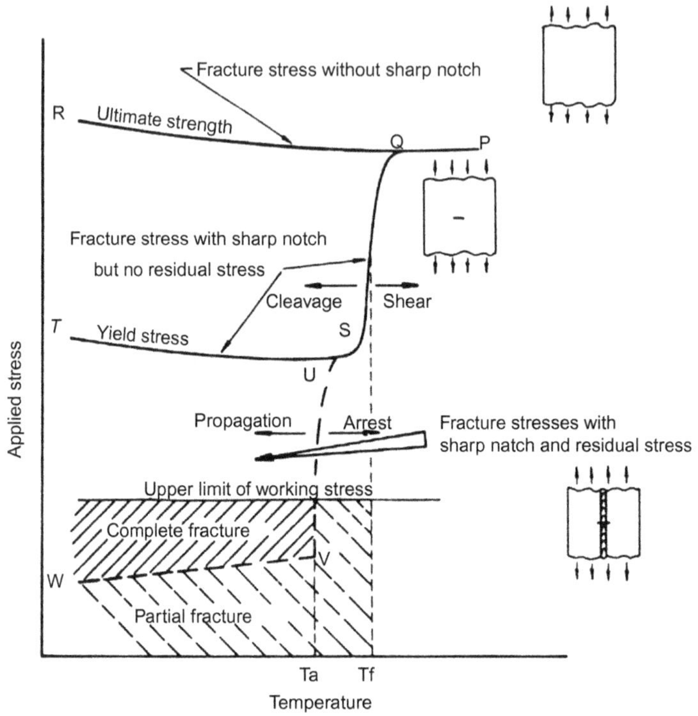

Figure 4.2 Effect of residual stress on brittle fracture of steel [50].

Under compressive loading, structures may fail by buckling. Figure 4.3 has been included to illustrate the role of RS in buckling in an I-shaped column [49]. The material of such column has been assumed to be perfectly elastic–plastic and represented by OABC in Figure 4.3A. In the inset of Figure 4.3A are I-columns under different loading levels. As loading level is increased, the black region which represents yielded portion of the column under compressive loading increases. Figure 4.3A illustrates that plastic yielding of the I-column takes place just beyond the load level equal to σ_p. The point at which yielding takes place is denoted by 1 in Figure 4.3A. On further increase of load, more regions yield and it is shown by the black region of I-column number 2 in the inset of Figure 4.3A. The load is indicated by 2 on the stress–strain curve. When load level reaches point B, entire structure yields and at that moment effect of RS on buckling behavior vanishes. Figure 4.3B shows the total strain and load carrying stresses corresponding to I-beam 1 shown in the inset of Figure 4.3A. This illustration highlights the role of compressive RS in the premature

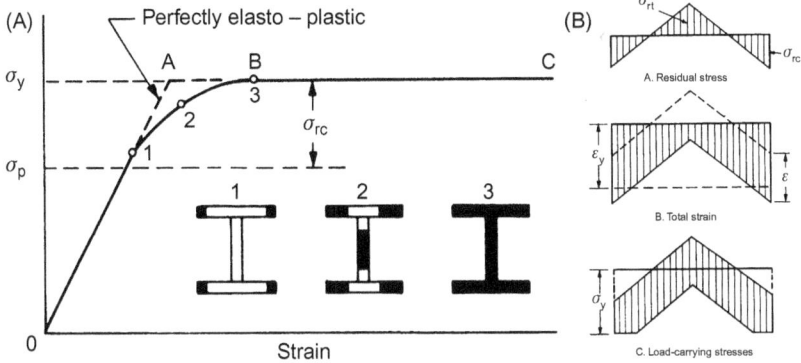

Figure 4.3 The effect of residual stress on buckling behavior of a I-column joint: (A) change of deformation behavior due to the presence of compressive residual stress and (B) residual stress distribution at different stages of compressive loading [49].

plastic deformation of structures, which may be undesirable in structural applications of such components.

Although static strength, fracture behavior, and buckling performance have been discussed based on the finding in fusion-welded components, the discussions are expected to be applicable for friction stir-welded structures also due to similarity of RS between fusion and FSW.

4.1.2 Fatigue

The understanding about the effect of RS on fatigue performance of FS-welded structures is no better than that for fusion-welded structures. As is the case with fusion-welded components, here also RS changes due to repeated loading. The change in RS due to repeated loading was discussed in Section 4.1.1. Since welding of any material is accompanied with microstructural changes, concomitant mechanical properties changes. Due to such metallurgical changes, the effect of RS on fatigue properties such as fatigue life and fatigue crack propagation (FCP) has not been ascertained.

Several studies have been reported so far on the influence of RS on cyclic deformation behavior of FS-welded structures. Although researchers have reported retardation of FCP rate in compressive RS zone and acceleration of FCP rate in tensile RS zone, it cannot be completely attributed to the RS. As mentioned earlier, the mechanical properties also change due to microstructural modification in welded zone and this can have significant influence on FCP rate. For example, tensile RS in the weld nugget is expected to enhance FCP rate but at the same time grain refinement in the nugget will also contribute toward enhanced FCP rate. Hence, it is very difficult to deconvolute the effect of these factors on FCP rate. Due to all these challenges, the effect of RS on fatigue performance of welded structures is inconclusive.

As discussed previously, cyclic loading may even out the residual stresses in welded components especially in low cycle fatigue regime. Hence, any change in fatigue behavior of the welded material in low cycle regime may be more related with the change in microstructure as a result of welding and less related to welding induced RS. However, one should exercise great caution in interpreting results while carrying out fatigue deformation behavior study in very high cycle regime where the strains are essentially elastic. Due to this, residual stress relaxation may not be possible owing to repeated loading and unloading. Hence, residual stress may play a significant role in influencing fatigue properties of welded materials in very high cycle regime.

4.2 STRESS-CORROSION CRACKING

Early work on the influence of RS on stress-corrosion cracking (SCC) by several researchers is mostly on fusion-welded steels. Omar [51] and Omar et al. [52] have studied SCC behavior of FS-welded 7075Al and 2195Al. Even after immersion of these welded samples in 3.5 wt% NaCl solution, no sign of SCC was observed. The conclusion was based on observation for cracks on the surface of the sample. However, SCC can lead to crack formation inside the sample which can be revealed by radiography as was the case for fusion butt-welded low-alloy high-strength steel immersed in a boiling aqueous solution of 60% $Ca(NO_3)_2$ and 4% NH_4NO_3 [53].

Due to very limited work carried out, it warrants a systematic work further to understand the role of RS on SCC behavior of FS weldments.

CHAPTER 5

Parameters Affecting RS

5.1 THE EFFECT OF TOOL TRAVERSE SPEED

It has been observed that RS and tool traverse speed are proportionally related, that is, RS increases with increase in tool traverse speed. The longer heating time of the material on the trailing side of the tool at lower tool traverse speed has been speculated to be a reason for such behavior. Other reason could be associated with relatively larger HAZ and larger drop in YS at low tool traverse speed due to higher heat input [25]. Peel et al. [22] processed 5083Al at three different tool traversed speed (100, 150, and 200 mm/min) to study the influence of traverse speed on residual stress. The results are shown in Figure 5.1. As can be noted, the value of peak longitudinal tensile stresses increased with increase in tool traverse speed. It also increased the longitudinal tensile stresses at the weld centerline. As mentioned before, it may be due to less thermal relief of a given volume of the material on the trailing side of the tool during welding. The transverse RS seem to remain unaffected with the change in traverse speed. The increasing traverse speed has impact on the width of the tensile zone also. With increase in traverse speed, a decrease in the tensile width in the welded region should be noted. Lombard et al. [24], Altenkirch et al. [12], Xu et al. [8], and Steuwer et al. [46] also have made similar observation for RS as a function of tool traverse speed.

The fourth set of the data points (filled square symbol) in Figure 5.1 has been included to show the effect of tool geometry on the residual stress. In this case, pin diameter was larger and pitch of the thread on the pin was coarser than the pin used for the rest of the welds. As we can see, hardly any effect of such change can be observed on the distribution of residual stress in the weldments.

5.2 THE EFFECT OF TOOL ROTATION RATE

The effect of tool rotation rate on RS have also been addressed by a few researchers [12,24]. Lombard et al. [24] used synchrotron X-ray

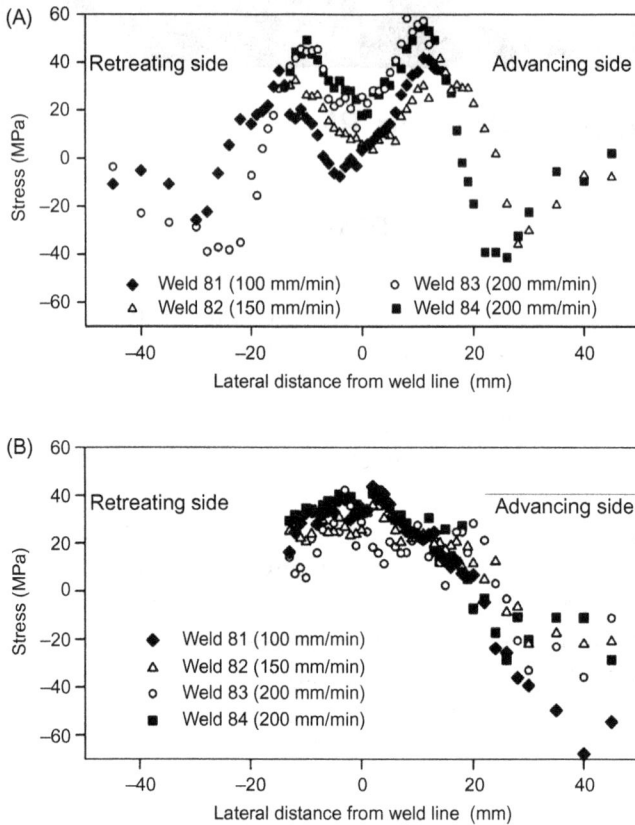

Figure 5.1 The effect of tool traverse speed on the RS: (A) longitudinal and (B) transverse, in 5083Al alloy measured using synchrotron XRD [22].

diffraction (XRD) technique to measure the RS in FS-welded 5083Al-H321 alloy. The result of the weld processed at tool traverse speed (feed rate 135 mm/min) and different rotation rates are shown in Figure 5.2. The outcome of this study is in consonant with tool traverse speed. With increase in tool rotation rate, a decrease in peak longitudinal tensile stress can be observed. In addition to this, overall the longitudinal tensile stresses at weld centerline are also decreasing with increase in tool rotation rate. It may be a consequence of thermal relief taking place at higher tool rotation rate. As speculated in Section 5.1 for lower tool traverse speed (higher heat input), at higher tool rotation rate (higher heat input), the drop in YS may be larger, which can cause lowering in RS with increase in tool rotation rate. The effect of

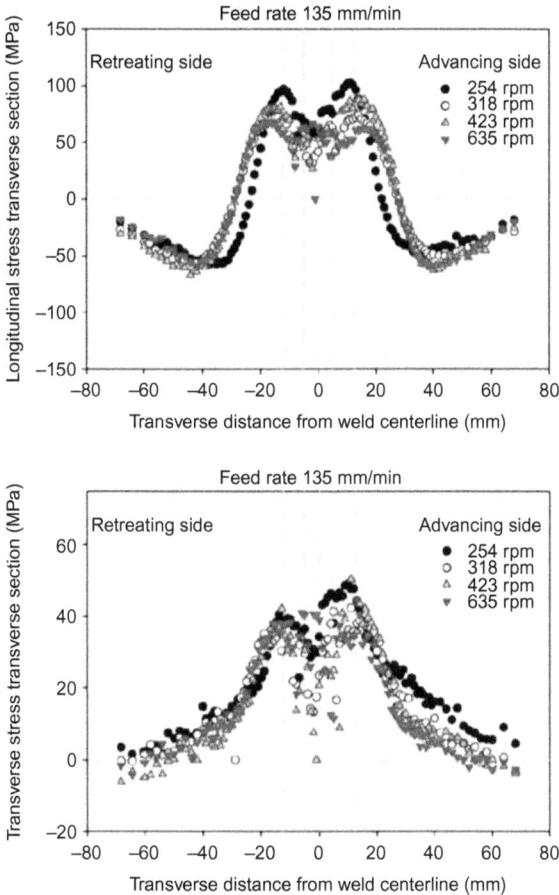

Figure 5.2 The effect of tool rotation rate on the RS in 5083Al-H321 alloy measured using synchrotron XRD [24].

increasing tool rotation rate at transverse RS appears to be less correlated as was the case for tool traverse speed. The width of the tensile zone is somewhat related with the magnitude of the tool rotation rate. As can be noted in Figure 5.2, at 423 rpm, the width of tensile zone is higher than that of at 254 rpm. However, no change on the tensile width can be observed on further increase in the tool rotation rate. Similar observations have been made by Altenkirch et al. [12] as far as the width of the tensile zone is concerned. However, in the work reported by Altenkirch et al. [12], hardly any influence of increase in tool rotation rate on the longitudinal RS can be observed unlike the one shown in Figure 5.2.

5.3 SAMPLE SIZE FOR MEASUREMENT OF RS

It is evident from Table 3.1 that significant difference exists in the sample dimensions used for welding. Very small to large sized samples have been used for making welds. In most of the cases, the samples have to be cut to smaller dimension either for RS measurement or evaluation of RS on various properties. This is done because in some cases, welded structure cannot be accommodated in the measurement device in its entirety or they cannot be tested for property evaluation in their original form. Cutting of welded samples may lead to relaxation of RS. Measurement made on such samples will lead to underestimation of RS which is not a desirable situation from design point of view.

Altenkirch et al. [15,31] have studied the effect of sectioning on RS relaxation in 7449Al alloy. The outcome of this study is shown in Figure 5.3. The relaxation of longitudinal RS as a function of fractional length across the width of the welded plate is clearly evident from this figure.

Figure 5.4 shows variation of longitudinal RS measured at the weld centerline as a function of fractional length of the plate. When fractional length of the welded plate is 0.4, very little relaxation can be noted in RS. At 10% fractional length, the RS is only 40% of the original RS. Altenkirch et al. [31] described the relaxation behavior of RS by the following exponential expression:

$$\sigma_{\text{relax}} = \sigma_0 \left[1 - \exp\left(-\frac{l_\text{r} - l_{\text{relax}}}{l_{\text{char}}}\right) \right] \qquad (5.1)$$

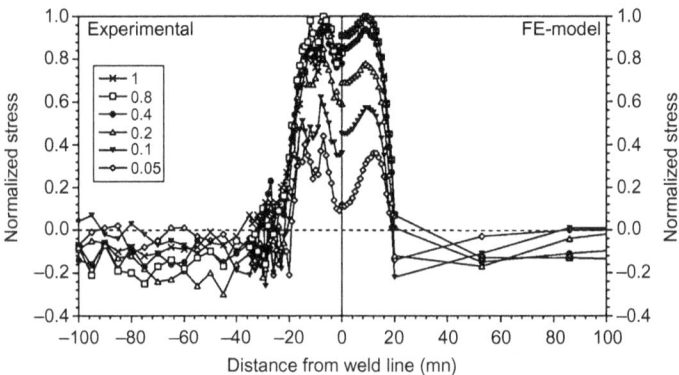

Figure 5.3 The effect of sectioning on RS relaxation in 7449Al alloy [31].

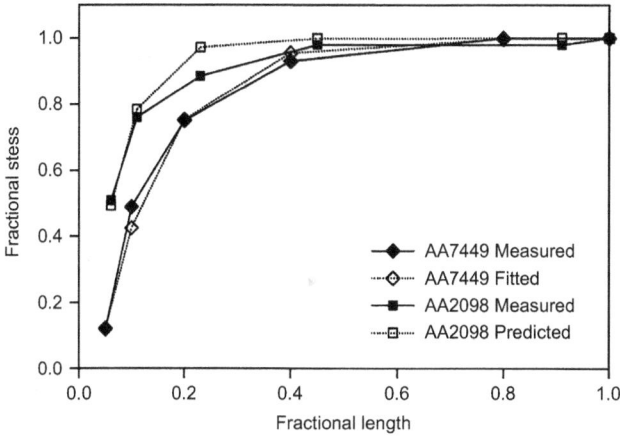

Figure 5.4 Variation of longitudinal RS at weld centerline as a function of fraction length of the 7449Al-welded plate [31].

where σ_{relax}, σ_0, l_r, l_{relax}, and l_{char} are stress after and before sectioning, remaining length, length at which RS becomes zero, and characteristics distance beyond which sectioning does not influence RS distribution, respectively. In fact, Eq. (5.1) is based on St. Venant's principle which defines a characteristics distance beyond which change in loading condition does not cause any change in state of stress at a given location in a structure [54]. In the work of Altenkirch et al. [15,31], l_{relax} and l_{char} were found to be equal to width of tensile RS zone (w) and three times of w. Hence, Eq. (5.1) can be rewritten as:

$$\sigma_{relax} = \sigma_0 \left[1 - \exp\left(\frac{1}{3} \left(1 - \frac{l_r}{w} \right) \right) \right] \qquad (5.2)$$

The review of literature data suggests that a wide variation in the sample size used for RS study exists which will have influence on the measured RS. Hence, the sample for RS measurement should be chosen such that it represents the RS of the weldments.

CHAPTER 6

Characterization of RS

From Table 3.1, it is evident that a wide range of techniques have been used to characterize RS in FS weldments. The most widely used techniques are cut-compliance, hole drilling, XRD, neutron diffraction, and synchrotron diffraction. Apart from them, ultrasonic wave technique and contour method have also been used for the characterization of RS. For a detailed discussion on each technique mentioned here, readers are referred to an overview article written by Withers and Bhadeshia [2] on the measurement technique for RS. However, a summary of these techniques have been provided in Table 6.1.

Table 6.1 Summary of Different Residual Stress Characterization Techniques and Corresponding Attributes

Method	Penetration	Spatial Resolution	Accuracy	Comments
Hole drilling (distortion caused by stress relaxation)	~1.2 × hole diameter	50 µm depth	±50 MPa, limited by reduced sensitivity with increasing depth	Measures in-plane type I stresses; semidestructive
Curvature (distortion as stresses arise or relax)	0.1–0.5 of thickness	0.05 of thickness; no lateral resolution	Limited by minimum measurable curvature	Unless used incrementally, stress field not uniquely determined; measures in-plane type I stresses
XRD (atomic strain gage)	<50 µm (Al); <5 µm (Ti); <1 mm with layer removal	1 mm laterally; 20 µm depth	±20 MPa, limited by nonlinearities in $\sin^2 \psi$ or surface condition	Nondestructive only as a surface technique; sensitive to surface preparation; peak shifts: types I, <II>; peak widths: types II, III
Hard X-rays (atomic strain gage)	150–50 mm (Al)	20 µm lateral to incident beam; 1 mm parallel to beam	±10 × 10^{-6} strain, limited by grain sampling statistics	Small gage volume leads to spotty powder patterns; peak shifts: types I, <II>, II; peak widths: types II, III
Neutrons (atomic strain gage)	200 mm (Al); 25 mm (Fe); 4 mm (Ti)	500 µm	±50 × 10^{-6} strain, limited by counting statistics and reliability of stress free references	Access difficulties; low data acquisition rate; costly; peak shifts: type I, <II> (widths rather broad)
Ultrasonics (stress-related changes in elastic wave velocity)	>10 cm	5 mm	10%	Microstructure sensitive; types I, II, III
Magnetic (variations in magnetic domains with stress)	10 mm	1 mm	10%	Microstructure sensitive; for magnetic materials only; types I, II, III
Raman	<1 µm	<1 µm approximately	$\Delta\lambda \approx 0.1$ cm$^{-1} \equiv 50$ MPa	Types I, II

Source: Reproduced from Ref. [2].

As can be noted from Table 3.1, depending on the technique employed, the values and trend shown by the individual components of the residual stress will vary significantly. Hence not only the size of the sample but also the measurement technique is very important for reliable evaluation of RS in weldments.

CHAPTER 7

Model for Understanding Residual Stress Development in Friction Stir-Welded Structures

7.1 DESCRIPTION OF THE MODEL

Figure 7.1 describes the development of RS during FSW process. Rectangular box A shows two regions: first, region II which is the region to be welded and second, region I which is the surrounding material. Region II has been again separately shown in Figure 7.1A as a small gray colored rectangular box. Here gray color represents room temperature indicated by T_0. During welding, the region II will attain a peak temperature which has been indicated by T_1 and it has been shown by red color now in Figure 7.1B. At this moment, we would like to invoke two situations:

- Case I—where there are no constraints at the sides of region II
- Case II—where there are rigid constraints (no deformation of constraints possible).

Case I has been represented in Figure 7.1D where it is free to expand as a consequence of thermal expansion. In such a situation, the change in dimension of the rectangle can be calculated using standard relationship. Let us assume that the thermal expansion can lead to a strain equal to C at this temperature (Figure 7.1D).

Case II is represented in Figure 7.1C. In this case, the region II is not allowed to expand perpendicular to the rigid clamps at the ends. To keep discussion simple, we'll assume that the region II cannot expand parallel to the surface of the rigid clamp (in reality it does). Since, region II now cannot expand due to the presence of rigid clamps at the end, the thermal strain will be zero. However, due to the presence of such a constraint, a thermal stress will be induced in region II. This will lead to introduction of a compressive mechanical strain C. The reason for mechanical strain to be equal to the thermal strain is as follows: if we assume that the box is free to expand as in Figure 7.1D,

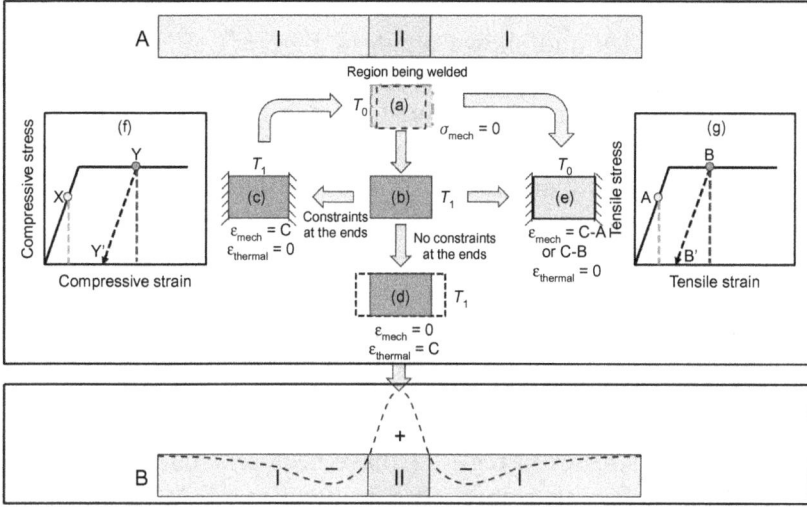

Figure 7.1 Schematic illustration of residual stress development during FSW or in general during solid-state joining process.

it will lead to a thermal strain C. To bring it to the original dimension, a compressive strain of equal amount will be needed to be applied. Since in Figure 7.1C, the thermal heating does not lead to expansion of the box, it is equivalent to the thermal expansion and mechanical compression mentioned above. The mechanical strain C can be equal to either strain X or Y shown in the schematic compressive stress–strain curve in Figure 7.1F. If during thermal heating compressive strain is equal to X which is elastic in nature, on cooling, there will be no dimensional change of the region II. However, if mechanical strain C reaches Y in that case, the region will plastically deform and it will try to shrink on cooling. Y and Y' represent total and plastic strains, respectively.

When region II comes down to room temperature T_0 from T_1, it will change its dimension depending on the state of strain. As mentioned before, if the compressive strain was elastic in strain, then there will be no change in the dimension (even if rigid constraints are present) otherwise it will shrink and it will be equal to Y' if the rigid clamps are removed. This shrinkage is represented by dashed rectangular box in Figure 7.1A. However, the complete shrinkage of region up on cooling will not take place due to constraints surrounding it as shown in rectangular box A by region I. For a rigid constraint (which does not deform elastically or plastically) on cooling, there will be no dimensional change of region II and it is shown by red dash-dotted box in Figure 7.1A and shown again in Figure 7.1E with rigid constraints. It is equivalent to the stretching of region II from the position where it was allowed to shrink by strain C to position where it does not shrink due to the presence of rigid constraints. It means the region II will be stretched by amount C. It will lead to tensile stresses in the region II. The amount of tensile stress will depend on strain C. If it leads to point A in Figure 7.1G, the region II will elastically deform. If the strain C leads to stretching to point B, then region II will plastically deform. In fact plastic deformation will be beneficial. It will lead to a reduction of total shrinkage required by the region II and consequently the tensile stresses.

These tensile stresses in region II have been schematically shown in rectangular box B with respect to consideration of A. In box A, the region I which acts as constraint for region II can deform elastically or plastically depending on the exact interaction between regions I and II. In the absence of all the external loads, the net force in a body is zero. Hence, to counter tensile stress in region II, region I develops compressive stresses in the vicinity of region II. Hence, box B shows a typical variation of RS in welded structures. In the next section, a validation of assumption made in this section will be provided using finite element analysis (FEA).

7.2 COMPUTATIONAL VALIDATION OF THE MODEL

The hypothesis put forth in the preceding section was validated using a two-dimensional (2D) finite element model. The model is schematically shown in Figure 7.2. The dimensions are given in the figure. The entire modeling was carried out in three steps—heating cycle: 10 s, cooling cycle: 500 s, and constraint removal: 10 s. Constraints are shown on the

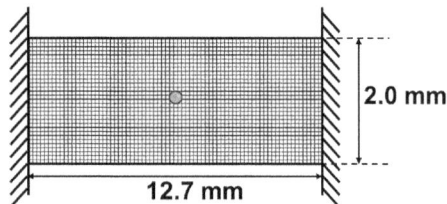

Figure 7.2 A schematic of the model used to evaluate model schematically presented in Figure 7.1.

right and left vertical edges of the model in Figure 7.2. A temperature of 250°C and 350°C was applied to the entire rectangular-shaped model enclosed within the constraints shown in the Figure 7.2 in two different analyses. It was applied for 10 s. It was followed by cooling cycle which lasted for 500 s. Such a large time was provided to cool the model sheet to close to room temperature in constrained condition. At the end of cooling cycle, the constraints were removed and the model was relaxed for 10 s.

The output of this analysis is shown in Figure 7.3. It shows the shrinkage of the sheet as a result of thermal heating at 250°C (Figure 7.3A) and 350°C (Figure 7.3B). The legend of the plot is shown in Figure 7.3C. As expected, the 2D model heated at 350°C showed higher shrinkage than the model heated at 250°C. This can be made out from the relative displacement between solid lines and dotted lines as shown in Figure 7.3A and B. Dotted lines represent undeformed shape, whereas solid lines in the figure represent deformed shape. To emphasize this point further, the regions enclosed by rectangular box (red, dotted) have been shown in Figure 7.3D and E. Here, the higher relative displacement between solid and dotted lines at

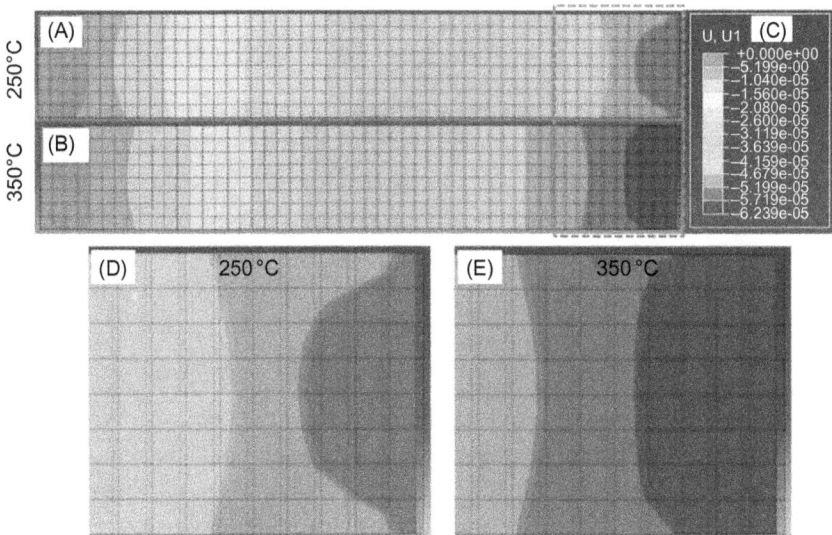

Figure 7.3 2D finite element model illustrating the shrinkage of the model as a result of heating at two different temperatures; (C) legend (D) and (E) are the enlarged views of the regions enclosed by red-dotted rectangular box as shown in (A) and (B).

350°C than that at 250°C is more evident. The maximum shrinkages at 250°C and 350°C were 5.55×10^{-5} and 6.24×10^{-5} m, respectively. Another point which is worth illustrating is the relative displacement along vertical direction. The dotted and solid lines in both the cases completely coincide along vertical direction which is indicative of very little or no deformation along vertical direction. Hence, the idea put forth in Section 7.1 that the constraints surrounding weld material leads to mechanical compressive strain is quite evident here.

Figure 7.4 shows the evolution of elastic strains, plastic strains, and stresses as a function of time at two different temperature levels—250°C and 350°C. Figure 7.4A, C, and E don't give clear picture about what is happening during heating stage due to very large scale of time axis. Hence, elastic strains, plastic strains, and stresses have been plotted up to 20 s (including 10 s of cooling stage) and shown in Figure 7.4B, D, and F, respectively. These measurements were taken at the location shown by red circular region (which is at the center of the rectangular region) in Figure 7.2. The salient points noted from Figure 7.4 are summarized in Table 7.1. The observations summarized in Table 7.1 are in line with the model discussed in Section 7.1.

7.3 DIFFERENCES IN RS IN FUSION WELDING AND FSW

Figure 7.5 has been included to highlight possible differences between residual stress development mechanism(s) between conventional fusion welding and FSW. Figure 7.5 shows the thermal stresses (von-Mises, S11, S22, YS as a function of temperature), equivalent plastic strain, and temperature as a function of time at weld centerline. The results were outcome of a finite element simulation study carried out for a 7050Al alloy.

Figure 7.5 shows that plastic deformation during FSW due to thermal excursion begins once YS of the alloy becomes equal to von-Mises stress (thermal stress). This point is labeled by numeral 1 next to equivalent plastic strain line near time axis in Figure 7.5A. To illustrate this point more clearly, YS, Mises stress, and equivalent plastic strain have been replotted again in Figure 7.5B. It should be noted that YS is a function of temperature, and temperature values corresponding to each YS value in Figure 7.5B should be read from Figure 7.5A by equating

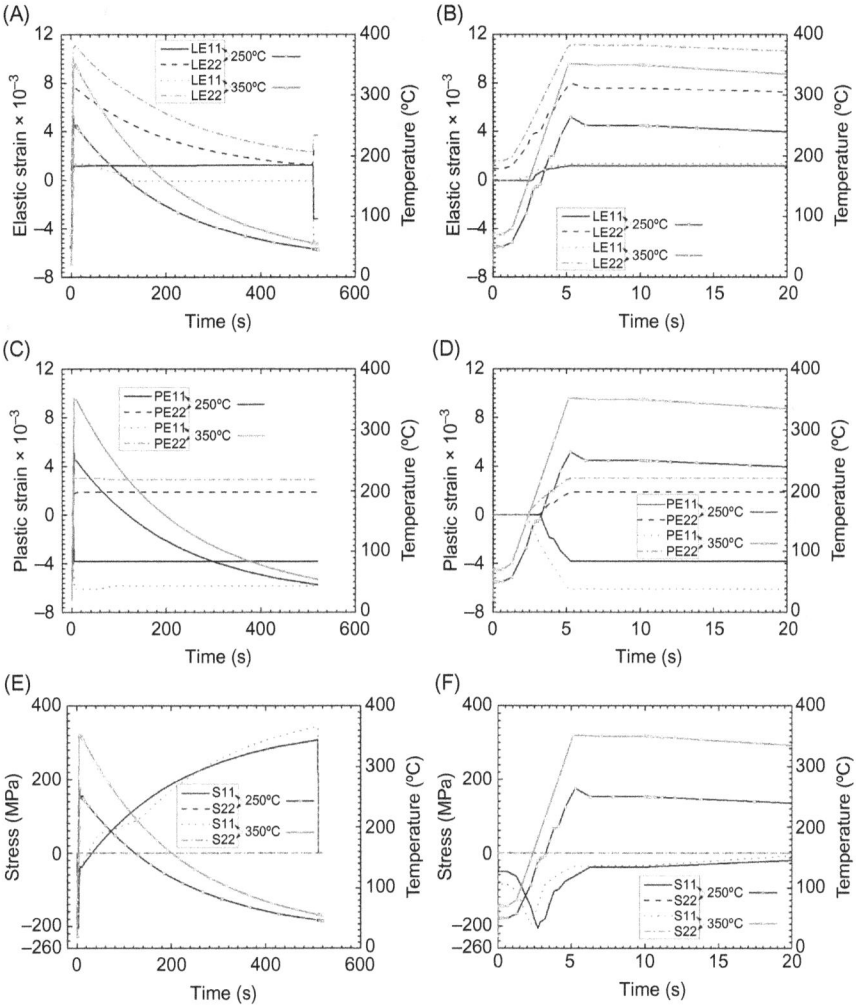

Figure 7.4 The plot showing the evolution of elastic strains (A) and (B), plastic strain (C) and (D), and stresses (E) and (F) as a function of time. Figures (B), (D), and (F) are enlarged plots (shown only up to 20 s) of figures (A), (C), and (E), respectively, for better visualization of the evolution of these parameters during heating stage of the welding. Heating cycle: 10 s, cooling cycle: 500 s, constraint or clamp removal: 10 s.

the corresponding time values. From Figure 7.5B, it is evident that the equivalent plastic strain starts rising upon the onset of the plastic deformation of the nugget. The rise in temperature further causes plastic deformation to proceed and hence rise in plastic strain curve. The rise of equivalent plastic strain curve continues during cooling cycle also unless temperature dropped below a particular value. During

Table 7.1 Summary of Key Observations Made in Figure 7.4
Elastic Strains (LE11 and LE22)
• LE11 < LE22 at both temperatures,   • LE22 approaches LE11 and LE11 = LE22 at the end of cooling cycle for 250°C. However, LE22 approaches LE11 but remains higher than LE11 at the end of cooling cycle for 350°C,   • On removing constraints, LE11 becomes compressive and LE22 more tensile for 250°C and 350°C as well,   • At higher temperature LE22 was higher than that at lower temperature. However, LE11 at 250°C and LE11 at 350°C remained same in magnitude for up to 40 s after which LE11 at 350°C decreased to a value slightly below zero and then progressively increased in magnitude (not evident from the plot).
Plastic Strains (PE11 and PE22)
• PE11 is compressive whereas PE22 is tensile in nature at both temperatures,   • In magnitude, PE11 is higher than PE22,   • PE11 and PE22 don't change with change in time for 250°C case,   • With increase in temperature, PE11 became more compressive and PE22 more tensile,   • PE11 and PE22 at 250°C remained constant throughout. However, PE11 and PE22 at 350°C decreased slightly in magnitude during cooling. The drop in PE11 and PE22 is indicative of tensile plastic strain due to stretching of the rectangular region.
Stresses (S11 and S22)
• During heating cycle, S11 at both temperatures become highly compressive and go through a minimum before becoming tensile in nature during cooling cycle. It attains its maximum value at the end of cooling cycle.   • S22 remained zero at both the temperature.   • S11 at 350°C was slightly higher than S11 at 250°C at the end of cooling cycle.   • On removing constraint. S11 became zero at both temperatures.   • At 250°C, during cooling S11 rises continuously whereas at 350°C there is a monotonic rise in the value of S11 during cooling but with a decrease in rate of increase in S11 for a small period of time. This is probably due to drop in LE11 and PE11.

cooling cycle when thermal stresses become smaller than YS of the material at that temperature, the plastic deformation of the nugget stops. At this point, the equivalent plastic strain reaches a plateau.

During fusion welding, the strain accumulated is relaxed completely once material goes to molten state. In the case of fusion welding, there are, mainly, two sources which lead to generation of RS—solidification shrinkage of the melt pool and plastic compressive strain of the solidified material similar to the one observed for FSW during cooling cycle. Hence, the RS are not much affected with the heating cycle. However, in FSW, the temperature never exceeds the melting point of the alloy; it is greatly influenced by heating cycle. As mentioned before RS during fusion welding develop from two sources—dimensional change of the material volume due to solidification shrinkage and mechanical compressive strain due to solidified welded material's interaction with surrounding material during cooling cycle. In FSW, although there is

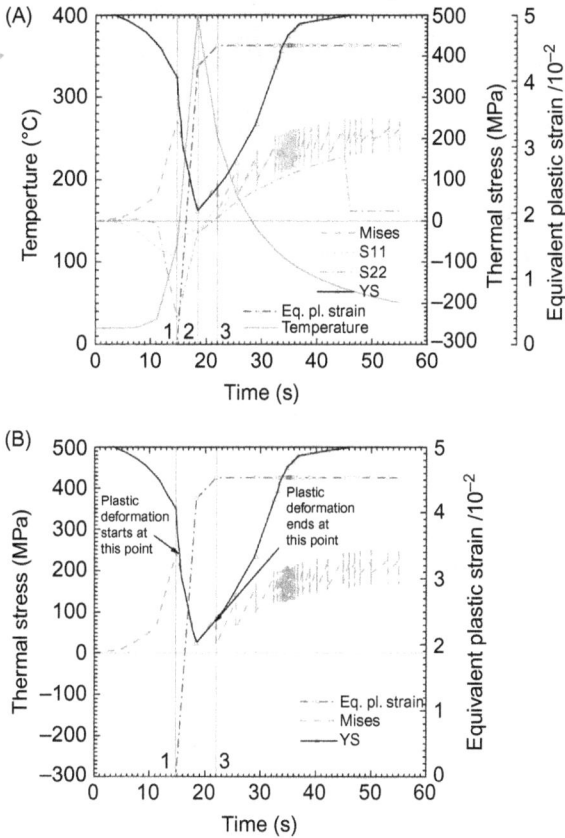

Figure 7.5 *(A) The temporal evolution of temperature, stress, strain, and YS at a point located at middle surface and weld centerline, halfway along the length of the plate; weld power = 2394 W (400 rpm/12 ipm); material properties temperature dependent; h_1 = 5000 W/m²-K and h_2 = 700 W/m²-K. (B) For better clarity only, Mises stress, equivalent plastic strain, and YS at different temperature have been shown. The temperature values should be read from Figure 7.5A.*

no contribution from solidification shrinkage due to obvious reasons, the compressive plastic strain accumulated during heating and cooling cycle leads to the overall development of RS. Due to this, the magnitude of RS are in some cases of friction stir-welded components comparable to that in fusion-welded components even though peak temperatures are significantly different.

CHAPTER 8

Mitigation of RS During FSW

8.1 THE EFFECT OF HEAT SINK (THERMAL TENSIONING) ON THE RESIDUAL STRESS

The effect of active cooling during FSW to mitigate residual stress and resulting distortion has been addressed by Staron et al. [55], Richards et al. [56], and Han et al. [11]. Significant reduction in RS as a result of *in situ* cooling has been reported. An example of the effect of *in situ* cooling is shown in Figure 8.1. As can be noted from Figure 8.1A, in absence of any active cooling, the longitudinal RS are tensile in the region around weld centerline. Figure 8.1B illustrates the effect of application of liquid CO_2 on the state of RS. Two nozzles directed liquid CO_2 at a distance 20 and 30 mm on the trailing side of the tool at the weld centerline. As can be noted, in the welded region, longitudinal tensile stresses have become compressive in nature. At the same time, no change in terms of nature and magnitude for longitudinal stresses in HAZ region can be observed. It has been attributed to ineffective cooling in HAZ region. Also, the nozzle distance beyond 30 mm did not provide any change in the state of residual stress. The work of Han et al. [11] also supports the above finding. A mixture of water and compressed air mixed by an atomizer were used on the trailing side of the tool to provide active cooling during FSW. Two different alloys—2024Al-T3 and 5083Al-H321—were FS-welded and in each case the effect of active cooling was significant. In the case of 2024Al-T3, the maximum tensile longitudinal residual stress decreased from 140 to ~50 MPa. Similarly, for 5083Al-H321, it dropped from 155 to ~65 MPa. Clearly, the use of *in situ* cooling medium is an effective way of controlling the RS in the weldments.

The modeling work of Richards et al. [56] also very vividly showed the impact of the size, power, and positioning of such cooling sources on the RS. A reduction as large as 100 MPa in RS at weld centerline was observed for 2024Al-T3 plates for a nozzle positioned at 40 mm on trailing side from the welding tool has been reported by these researchers.

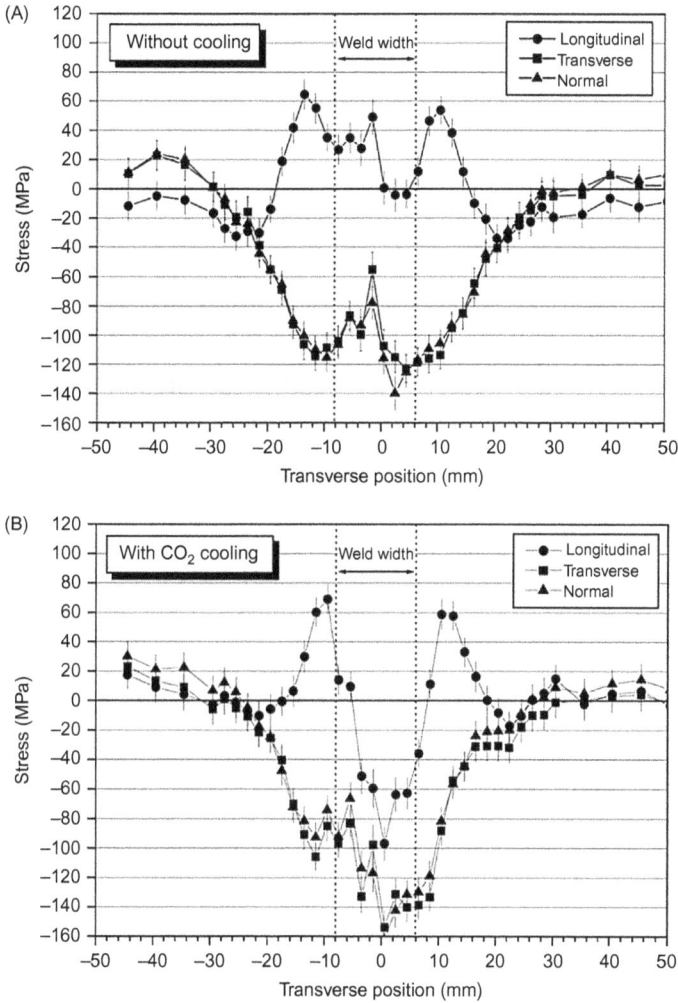

Figure 8.1 The effect of in situ liquid CO_2 cooling on trailing side of the tool during FSW of 2024Al-T351; residual stress distribution (A) without cooling and (B) with cooling [55].

8.2 THE EFFECT OF MECHANICAL TENSIONING ON THE RESIDUAL STRESS

As the name implies, the plates or sheets being welded are clamped at one end and pulled from other end of the plate and the direction of pull being along the weld direction. This has been very extensively used in other welding processes and has been adopted by FSW community to mitigate the problem of RS and resulting distortions.

Staron et al. [57] applied mechanical tensioning technique to investigate its impact of residual stress distribution. 2024Al-T351 was subjected to mechanical tensioning before FSW. The result of their work is illustrated in Figure 8.2. It is evident from this figure that on mechanical tensioning, both longitudinal and transverse RS became compressive in nature not only in the region around the weld centerline but in major part of the plate. Similar effect of mechanical tensioning on RS has been observed by other researchers also [20,31,58].

8.3 THE EFFECT OF ROLLER TENSIONING ON THE RESIDUAL STRESS

The roller tensioning can be considered as a variant of mechanical tensioning. Instead of applying the load for stretching the plate or sheet being welded at one end, in this case roller is moved over the welded region on the trailing side or in regions next to FS weld. It has been schematically shown in Figure 8.3.

This technique was used by Altenkirch et al. [14] to investigate its effects on residual stress distribution. The roller shown in Figure 8.3A was employed on the trailing side of the tool while FSW was going on (*in situ* roller tensioning). The other roller (Figure 8.3B) was positioned directly over the weld after the FSW was over (post-weld roller tensioning). The result of their work is shown in Figure 8.4. Very little or no effect can be observed on RS of *in situ* roller tensioning

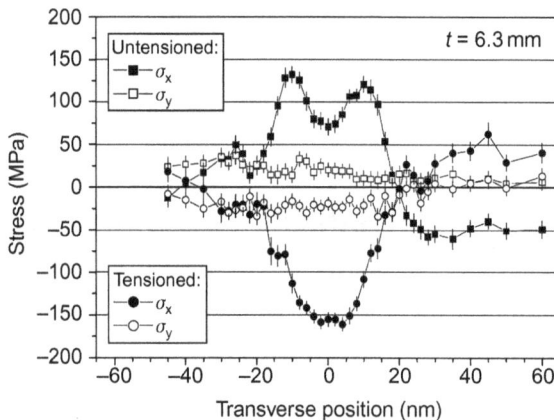

Figure 8.2 The effect of mechanical tensioning (tensile stress 70% of the room temperature YS of the plate) on FS-welded 2024Al-T351 [57].

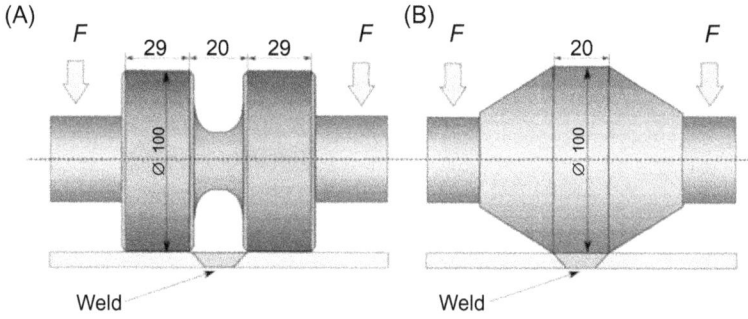

Figure 8.3 Schematic representation of rollers and their arrangements for roller tensioning in FSW to mitigate RS [14].

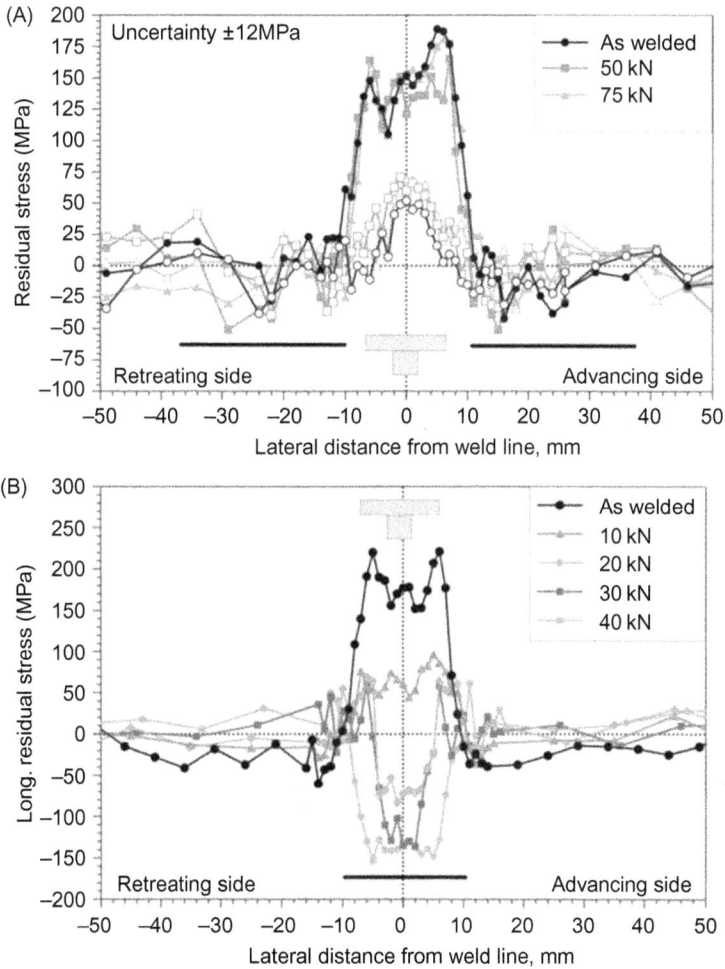

Figure 8.4 The effect of roller tensioning on the residual stress distribution: (A) in situ roller tensioning with the roller shown in Figure 8.3A and (B) post-weld roller tensioning with the roller shown in Figure 8.3B [14].

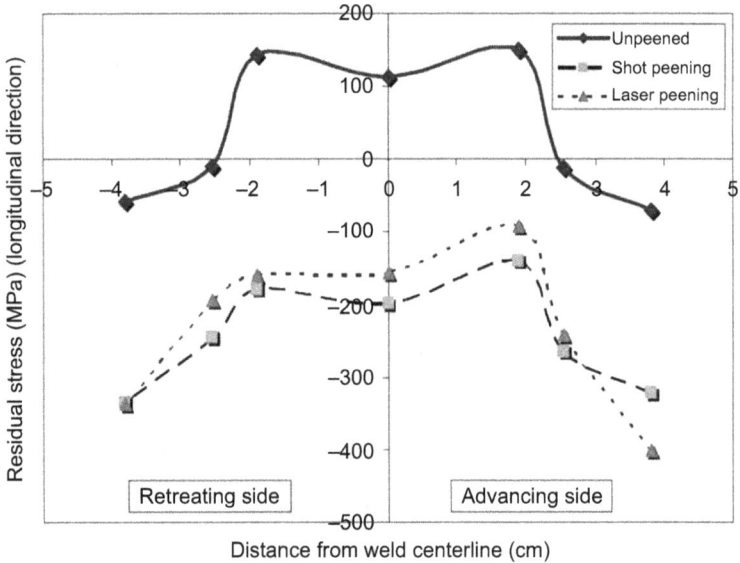

Figure 8.5 The effect of laser and shot peening on the residual stress distribution in 2195Al alloy [17].

(Figure 8.4A). Figure 8.4B shows that the nature of RS completely reversed on application of post-weld roller tensioning.

8.4 THE EFFECT OF LASER AND SHOT PEENING ON RS

Hatamleh [17] studied the effect of laser and shot peening of FS weldments in order to evaluate the state of RS in the weldments. As expected, the tensile longitudinal RS became compressive on laser and shot peening. Figure 8.5 summarizes the result of laser and shot peening carried out on FS-welded 2195Al. As can be seen, longitudinal RS became completely compressive as a result of laser and shot peening.

As a consequence of laser and shot peening, the surface of the material gets plastically deformed and tries to expand. However, the material lying below surface prevents it from expanding which leads to introduction of compressive stresses in the surface.

CHAPTER 9

Simulation of FSW for RS and Distortions

The overview of experimental work carried out so far to understand RS in FS-welded structures suggests that there are several drawbacks to the present approach:

1. There is a lack of systematic study to understand RS evolution in FS weldments. A wide variation exists in the sheet/plate dimensions used for RS study. It makes comparison of results published by various researchers very difficult and general conclusions cannot be drawn due to this.
2. Different techniques have different attributes. Hence, with other variables constant, the deployment of different measurement techniques would lead to differences in trend and magnitude of RS.
3. Sectioning of welded structures for RS measurement leads to stress relaxation. If such relaxation is not accounted for, it may lead to a faulty design. Experimentally stress relaxation can be assessed at few locations only. It is quite a daunting task to evaluate stress relaxation experimentally for the entire welded structures.
4. Most of the experimental works focus on stress distribution measurement at single location (at single cross-section) in the entire weldments. However, limited experimental work has shown that nature and magnitude of RS distribution may significantly change from one end of the plate to the other end. Also, a considerable variation in through thickness RS may exist in thick plates. A thorough evaluation of RS experimentally in the entire weldments is very time consuming, costly, and painstaking task.

In view of above observations, computational tools are becoming indispensable for studying development of RS in FSW [59,60]. There are several reasons for ever-increasing use of computational techniques for understanding evolution of RS in weldments:

1. With advancement in computer technology, even desktop computers are fast enough to analyze a sufficiently large problem in a reasonable period of time.

2. It is possible to analyze 3D geometries with all essential features of welding required for accurate prediction of RS.
3. It is possible to study the effect of various welding parameters on the evolution of thermo-mechanical stresses during welding and eventually evolution of RS.
4. There are several computational fluid dynamics and continuum FEA commercial packages available today which make computational analysis of such problems very easy.

In a recent special issue on RS in welding, De and DebRoy [59] have listed three major challenges faced by computer-based numerical modeling: (1) need for 3D analysis incorporating conduction and convection heat transfer modes for accurate computation of thermal field, (2) thermo-mechanical evolution of plastic strain as a function of transient temperature field and complex mechanical constraints, and (3) need for correct constitutive models for predicting thermo-elasto-plastic material behavior.

The consideration of convective heat transfer becomes important in the modeling of thermal field and physical shape of the weld pool in fusion welding processes. However, research has shown that even in fusion welding processes, mere consideration of conduction mode of heat transfer is enough for the prediction of RS and distortions. Given that there is no melting during FSW process, just consideration of conduction heat transfer should suffice the simulation of RS and resulting distortions. In FSW community, two different approaches have been taken so far in modeling thermal history of the process—pure conduction mode-based and computational fluid dynamics (CFD)-based approaches. The thermal history predicted by these models is then distributed in a thermo-elasto-plastic-based Lagrangian model for the prediction of RS and distortions.

Zhu and Chao [61] carried out numerical simulation of RS in FSW of 304L stainless steel. It was simulated using a code developed specifically for welding called WELDSIM. In this model, two layers of eight noded brick (hexahedral) elements were used in thickness direction. The dimension of the modeled plate was 304.8 mm × 203.2 mm × 3.18 mm. A graded meshing strategy was used in which mesh density gradually increased on moving toward the weld centerline from outer edge of the plate. The RS before and after fixture release was calculated. A significant decrease in RS (longitudinal) was observed on release of fixture.

Also, the predicted residual stress after release of the fixture was in close agreement with the experimentally reported values.

Zhang et al. [61] studied evolution of residual stress during FSW process by developing a 2D FSW model. To incorporate large plastic deformation around tool, arbitrary Lagrangian−Eulerian formulation and adaptive remeshing-based model was developed. The model was meshed using four noded quadrilateral elements. The material was modeled as rate-independent elastic−plastic material and the model was able to predict the trend and values generally observed for 6061Al-T6 experimentally.

In all the earlier work, the focus was on predicting RS as a function of various welding variables [61−66]. However, now trend is moving toward understanding processes which can reduce RS. For example, the experimental work carried out by Staron et al. [55] showed a significant reduction in tensile stresses (became compressive) in the weld nugget due to active cooling on the trailing side of tool (Figure 7.5). Richards et al. [56] have studied this aspect computationally and showed variation in RS by changing the position of nozzle used for cooling with respect to the tool. Figure 9.1 includes the results of the work done by Richards et al. [56] for mitigating RS using active cooling. It shows different positions of the cooling nozzle. Figure 9.1E shows the use of two nozzles in such studies. Figure 9.2 shows the distribution of longitudinal residual stress across the width of welded structure at mid-thickness of the plate. As can be noted, the positioning of cooling nozzle with respect to the tool greatly affects the

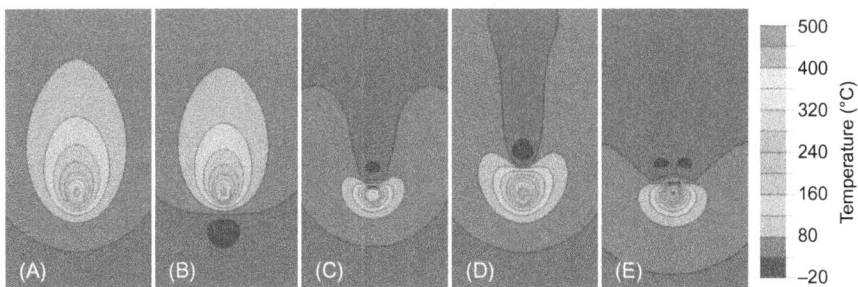

Figure 9.1 Temperature contour plot as a function of distance of cooling nozzle from heat source: (A) no cooling, (B) 20 mm in front of the heat source; in rest of the cases heat cooling source behind the heat source, (C) 20 mm, (D) 40 mm, and (E) two heat sinks at 20 mm [56].

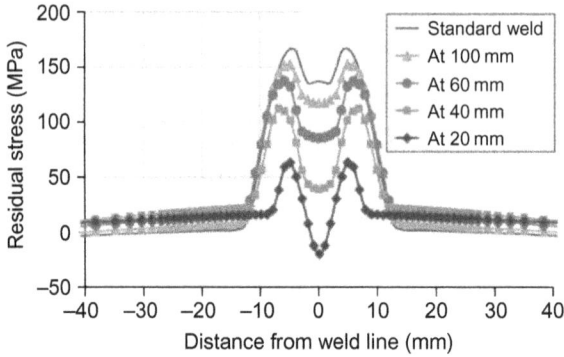

Figure 9.2 Effect of active cooling and the position of the cooling medium on the longitudinal residual stress distribution. The distribution is at mid-thickness level halfway along the length of the plate [56].

magnitude of RS. Such studies can be used as guide to experimental work designed to mitigate RS.

The present-day models are even able to consider mechanical property distribution due to change in microstructure during welding [67]. Modeling effort has gone even toward understanding how plate sectioning can result into relaxation of RS in the welded structures [68].

Hence, a brief review of the state of RS study in FSW using computational tools suggests that a considerable progress has been made so far and efforts should be directed toward understanding the mechanisms of RS generation as a function of process variables. At present, continuum-based FEA tools can't handle large plastic deformation associated with FSW process. Therefore, at present, the general trend is to consider plastic deformation of the material due to thermal excursion only. However, improved FEA codes coupled with suitable material model would lead to better prediction of RS and consequently improved understanding of their origin. It will help deployment of RS and distortion mitigation tools in a much more efficient manner.

CHAPTER 10

Summary, Conclusions, and Future Direction

The knowledge of RS study is very important from the point of intended performance and service life of the components. At present, a great deal of research has gone into understanding the origin of RS either experimentally or computationally. It has helped improve our understanding regarding sources of RS in the weldments. Extensive research carried out on RS also provides us with a guideline in mitigating the RS and resulting distortion. Most of the research carried out have focused on aluminum alloys. There is a need to extend the research toward other alloys as application of FSW becomes more widespread. Also, RS research for ferrous and other high-strength alloys will become necessary as FSW of these alloys becomes economically viable.

There is still some gap in our knowledge for improving the understanding of RS. There is no extensive study at present to elucidate the effect of plate dimension on RS. Other processing parameters can also be varied systematically in order to study their influence on the RS. Also, there is a need to investigate the role of RS measurement techniques on the values and trend reported in the literature. Simulation of RS is a great tool to visualize thermo-mechanical stresses in the weldments at different stages of welding. However, simulation of FSW process is still at developmental stage due to huge computational cost associated with modeling the complexity involved in FSW. Although with simplifying assumptions present models are able to predict the trend of RS, accuracy of values predicted will depend on the constitutive material model used. The development of constitutive material model can be another area of active research.

REFERENCES

[1] T. Inoue, Z. Wang, Coupling between stress, temperature, and metallic structures during processes involving phase transformations, Mater. Sci. Technol. 1 (1985) 845–850.

[2] P. Withers, H. Bhadeshia, Residual stress part 1—measurement techniques, Mater. Sci. Technol. 17 (2001) 355–365.

[3] A. Leon, Benefits of split mandrel coldworking, Int. J. Fatigue 20 (1998) 1–8.

[4] L. Wagner, Mechanical surface treatments on titanium, aluminum and magnesium alloys, Mater. Sci. Eng. A. Struct. Mater. 263 (1999) 210–216.

[5] W.M. Thomas, E.D. Nicholas, J.C. Needham, M.G. Murch, P.T. Smith, C.J. Dawes, Friction welding, US Patent 5,460,317 (1995).

[6] R.S. Mishra, Friction stir processing, in: R.S. Mishra, M.W. Mahoney (Eds.), Friction Stir Welding and Processing, ASM International, Materials Park, OH, 2007, pp. 309–314.

[7] R.S. Mishra, Z.Y. Ma, Friction stir welding and processing, Mater. Sci. Eng. R 50 (2005) 1–78.

[8] W. Xu, J. Liu, H. Zhu, Analysis of residual stresses in thick aluminum friction stir welded butt joints, Mater. Des. 32 (2011) 2000–2005.

[9] L. Wang, C.M. Davies, R.C. Wimpory, L.Y. Xie, K.M. Nikbin, Measurement and simulation of temperature and residual stress distributions from friction stir welding AA2024 Al alloy, Mater. High Temperatures 27 (2010) 167–178.

[10] J.R. Davis, ASM International, Aluminum and Aluminum Alloys, ASM International, Materials Park, OH, 1993.

[11] W.T. Han, F.R. Wan, G. Li, C.L. Dong, J.H. Tong, Effect of trailing heat sink on residual stresses and welding distortion in friction stir welding Al sheets, Sci. Technol. Weld. Joining 16 (2011) 453–458.

[12] J. Altenkirch, A. Steuwer, P.J. Withers, Process–microstructure–property correlations in Al–Li AA2199 friction stir welds, Sci. Technol. Weld. Joining 15 (2010) 522–527.

[13] Y.E. Ma, P. Staron, T. Fischer, P.E. Irving, Size effects on residual stress and fatigue crack growth in friction stir welded 2195-T8 aluminium—Part I: experiments, Int. J. Fatigue 33 (2011) 1417–1425.

[14] J. Altenkirch, A. Steuwer, P.J. Withers, S.W. Williams, M. Poad, S.W. Wen, Residual stress engineering in friction stir welds by roller tensioning, Sci. Technol. Weld. Joining 14 (2009) 185–192.

[15] J. Altenkirch, A. Steuwer, M.J. Peel, P.J. Withers, The extent of relaxation of weld residual stresses on cutting out cross-weld test-pieces, Powder Diffr. 24 (2009) S31–S36.

[16] L. Fratini, S. Pasta, A.P. Reynolds, Fatigue crack growth in 2024-T351 friction stir welded joints: longitudinal residual stress and microstructural effects, Int. J. Fatigue 31 (2009) 495–500.

[17] O. Hatamleh, A comprehensive investigation on the effects of laser and shot peening on fatigue crack growth in friction stir welded AA 2195 joints, Int. J. Fatigue 31 (2009) 974–988.

[18] G. Pouget, A.P. Reynolds, Residual stress and microstructure effects on fatigue crack growth in AA2050 friction stir welds, Int. J. Fatigue 30 (2008) 463–472.

[19] T. Li, Q.Y. Shi, H.K. Li, W. Wang, Z.P. Cai, Residual stresses of friction stir welded 2024-T4 joints, Trans. Tech. Publ. 580 (2008) 263–266.

[20] D.A. Price, S.W. Williams, A. Wescott, C.J.C. Harrison, A. Rezai, A. Steuwer, et al., Distortion control in welding by mechanical tensioning, Sci. Technol. Weld. Joining 12 (2007) 620–633.

[21] H.J.K. Lemmen, R.C. Alderliesten, R.R.G.M. Pieters, R. Benedictus, J.A. Pineault, Yield strength and residual stress measurements on friction-stir-welded aluminum alloys, J. Aircr. 47 (2010) 1570–1583.

[22] M. Peel, A. Steuwer, M. Preuss, P.J. Withers, Microstructure, mechanical properties and residual stresses as a function of welding speed in aluminium AA5083 friction stir welds, Acta Mater. 51 (2003) 4791–4801.

[23] M. James, D. Hughes, D. Hattingh, G. Bradley, G. Mills, P. Webster, Synchrotron diffraction measurement of residual stresses in friction stir welded 5383-H321 aluminium butt joints and their modification by fatigue cycling, Fatigue Fract. Eng. Mater. Struct. 27 (2004) 187–202.

[24] H. Lombard, D.G. Hattingh, A. Steuwer, M.N. James, Effect of process parameters on the residual stresses in AA5083-H321 friction stir welds, Mater. Sci. Eng. A Struct. Mater. 501 (2009) 119–124.

[25] X. Wang, Z. Feng, S. David, S. Spooner, C. Hubbard, Neutron diffraction study of residual stresses in friction stir welds, CRS-6 Proceedings (2000) 1408–1415.

[26] A. Steuwer, M.J. Peel, P.J. Withers, Dissimilar friction stir welds in AA5083-AA6082: the effect of process parameters on residual stress, Mater. Sci. Eng. A Struct. Mater. 441 (2006) 187–196.

[27] W. Woo, H. Choo, D.W. Brown, Z. Feng, P.K. Liaw, Angular distortion and through-thickness residual stress distribution in the friction-stir processed 6061-T6 aluminum alloy, Mater. Sci. Eng. A Struct. Mater. 437 (2006) 64–69.

[28] G. Buffa, L. Fratini, S. Pasta, R. Shivpuri, On the thermo-mechanical loads and the resultant residual stresses in friction stir processing operations, Cirp Ann. Manuf. Technol. 57 (2008) 287–290.

[29] S. Gachi, F. Belahcene, F. Boubenider, Residual stresses in AA7108 aluminium alloy sheets joined by friction stir welding, Nondestr. Test. Eval. 24 (2009) 301–309.

[30] R. Brown, W. Tang, A.P. Reynolds, Multi-pass friction stir welding in alloy 7050-T7451: effects on weld response variables and on weld properties, Mater. Sci. Eng. A Struct. Mater. 513–14 (2009) 115–121.

[31] J. Altenkirch, A. Steuwer, M. Peel, D.G. Richards, P.J. Withers, The effect of tensioning and sectioning on residual stresses in aluminium AA7749 friction stir welds, Mater. Sci. Eng. A Struct. Mater. 488 (2008) 16–24.

[32] L. Fratini, B. Zuccarello, An analysis of through-thickness residual stresses in aluminium FSW butt joints, Int. J. Mach. Tools Manufacture 46 (2006) 611–619.

[33] M. Sutton, A. Reynolds, D. Wang, C. Hubbard, A study of residual stresses and microstructure in 2024-T3 aluminum friction stir butt welds, J. Eng. Mat. Technol. Trans. Asme 124 (2002) 215–221.

[34] M. James, D. Hattingh, D. Hughes, L. Wei, E. Patterson, J. Da Fonseca, Synchrotron diffraction investigation of the distribution and influence of residual stresses in fatigue, Fatigue Fract. Eng. Mater. Struct. 27 (2004) 609–622.

[35] S. Hong, S. Kim, C.G. Lee, S. Kim, Fatigue crack propagation behavior of friction stir welded 5083-H32 Al alloy, J. Mater. Sci. 42 (2007) 9888–9893.

[36] K. Deplus, A. Simar, W. Van Haver, B. de Meester, Residual stresses in aluminium alloy friction stir welds, Int. J. Adv. Manuf. Technol. 56 (2011) 493–504.

[37] M.B. Prime, T. Gnaeupel-Herold, J.A. Baumann, R.J. Lederich, D.M. Bowden, R.J. Sebring, Residual stress measurements in a thick, dissimilar aluminum alloy friction stir weld, Acta Mater. 54 (2006) 4013–4021.

[38] H.J. Aval, S. Serajzadeh, N.A. Sakharova, A.H. Kokabi, A. Loureiro, A study on microstructures and residual stress distributions in dissimilar friction-stir welding of AA5086-AA6061, J. Mater. Sci. 47 (2012) 5428–5437.

[39] A. Scialpi, M. De Giorgi, L.A.C. De Filippis, R. Nobile, F.W. Panella, Mechanical analysis of ultra-thin friction stir welding joined sheets with dissimilar and similar materials, Mater. Des. 29 (2008) 928–936.

[40] W. Woo, H. Choo, M.B. Prime, Z. Feng, B. Clausen, Microstructure, texture and residual stress in a friction-stir-processed AZ31B magnesium alloy, Acta Mater. 56 (2008) 1701–1711.

[41] W. Woo, H. Choo, Softening behaviour of friction stir welded Al 6061-T6 and Mg AZ31B alloys, Sci. Technol. Weld. Joining 16 (2011) 267–272.

[42] S. Pasta, A.P. Reynolds, Evaluation of residual stresses during fatigue test in an FSW joint, Strain 44 (2008) 147–152.

[43] S. Pasta, A.P. Reynolds, Residual stress effects on fatigue crack growth in a Ti–6Al–4V friction stir weld, Fatigue Fract. Eng. Mater. Struct. 31 (2008) 569–580.

[44] Y. Lee, J.-. Kim, J.-. Lee, K. Kims, J.Y. Koo, D. Kwon, Using the instrumented indentation technique for stress characterization of friction stir-welded API X80 steel, Philos. Mag. 86 (2006) 5497–5504.

[45] M.H. Mathon, V. Klosek, Y. de Carlan, L. Forest, Study of PM2000 microstructure evolution following FSW process, J. Nucl. Mater. 386–88 (2009) 475–478.

[46] A. Steuwer, S.J. Barnes, J. Altenkirch, R. Johnson, P.J. Withers, Friction stir welding of HSLA-65 steel: part II. The influence of weld speed and tool material on the residual stress distribution and tool wear, Metall. Mater. Trans. A 43A (2012) 2356–2365.

[47] A.K. Lakshminarayanan, V. Balasubramanian, Assessment of fatigue life and crack growth resistance of friction stir welded AISI 409M ferritic stainless steel joints, Mater. Sci. Eng. A Struct. Mater. 539 (2012) 143–153.

[48] K. Masubuchi, The magnitude and distribution of residual stresses in weldments, in: D.W. Hopkins (Ed.), Analysis of Welded Structures, Pergamon Press, Great Britain, 1980, p. 189.

[49] K. Masubuchi, The strength of welded structure: fundamentals, in: D.W. Hopkins (Ed.), Analysis of Welded Structures, Pergamon Press, Great Britain, 1980, p. 328.

[50] H. Kihara, K. Masubuchi, Effect of residual stress on brittle fracture, Weld. J. 38 (1959) 159s–168s.

[51] O. Hatamleh, P.M. Singh, H. Garmestani, Corrosion susceptibility of peened friction stir welded 7075 aluminum alloy joints, Corros. Sci. 51 (2009) 135–143.

[52] O. Hatamleh, P.M. Singh, H. Garmestani, Stress corrosion cracking behavior of peened friction stir welded 2195 aluminum alloy joints, J. Mater. Eng. Perform. 18 (2009) 406–413.

[53] K. Masubuchi, The magnitude and distribution of residual stresses in weldments, in: D.W. Hopkins (Ed.), Analysis of Welded Structures, Pergamon Press, Great Britain, 1980, p. 478.

[54] S.P. Timoshenko, J. Goodier, Theory of Elasticity, 1970, McGraw-Hill, New York, 1987 (1970).

[55] P. Staron, M. Kocak, S. Williams, Residual stresses in friction stir welded Al sheets, Appl. Phys. A-Mater. 74 (2002) S1161−S1162.

[56] D.G. Richards, P.B. Prangnell, P.J. Withers, S.W. Williams, T. Nagy, S. Morgan, Efficacy of active cooling for controlling residual stresses in friction stir welds, Sci. Technol. Weld. Joining 15 (2010) 156−165.

[57] P. Staron, M. Kocak, S. Williams, A. Wescott, Residual stress in friction stir-welded Al sheets, Phys. B-Condensed Matter. 350 (2004) E491−E493.

[58] D.G. Richards, P.B. Prangnell, S.W. Williams, P.J. Withers, Global mechanical tensioning for the management of residual stresses in welds, Mater. Sci. Eng. A Struct. Mater. 489 (2008) 351−362.

[59] A. De, T. DebRoy, A perspective on residual stresses in welding, Sci. Technol. Weld. Joining 16 (2011) 204−208.

[60] P. Michaleris, Modelling welding residual stress and distortion: current and future research trends, Sci. Technol. Weld. Joining 16 (2011) 363−368.

[61] X. Zhu, Y. Chao, Numerical simulation of transient temperature and residual stresses in friction stir welding of 304L stainless steel, J. Mater. Process. Technol. 146 (2004) 263−272.

[62] H. Zhang, Z. Zhang, J. Chen, The finite element simulation of the friction stir welding process, Mater. Sci. Eng. A Struct. Mater. 403 (2005) 340−348.

[63] Y.J. Chao, X. Qi, Thermal and thermo-mechanical modeling of friction stir welding of aluminum alloy 6061-T6, J. Mater. Process. Manuf. Sci. 7 (1998) 215−233.

[64] C.M. Chen, R. Kovacevic, Parametric finite element analysis of stress evolution during friction stir welding, Proc. Inst. Mech. Eng. B-J. Eng. Manufacture 220 (2006) 1359−1371.

[65] M.Z.H. Khandkar, J.A. Khan, A.P. Reynolds, M.A. Sutton, Predicting residual thermal stresses in friction stir welded metals, J. Mater. Process. Technol. 174 (2006) 195−203.

[66] Q.-. Shi, J. Silvanus, Y. Liu, D.-. Yan, H.-. Li, Experimental study on distortion of Al-6013 plate after friction stir welding, Sci. Technol. Weld. Joining 13 (2008) 472−478.

[67] Z. Feng, X.-. Wang, S.A. David, P.S. Sklad, Modelling of residual stresses and property distributions in friction stir welds of aluminium alloy 6061-T6, Sci. Technol. Weld. Joining 12 (2007) 348−356.

[68] M. Law, O. Kirstein, V. Luzin, An assessment of the effect of cutting welded samples on residual stress measurements by chill modelling, J. Strain Anal. Eng. Des. 45 (2010) 567−573.